8th EASN-CEAS Workshop on Manufacturing for Growth and Innovation

8th EASN-CEAS Workshop on Manufacturing for Growth and Innovation

Special Issue Editors

Konstantinos Kontis
Spiros Pantelakis

MDPI • Basel • Beijing • Wuhan • Barcelona • Belgrade

Special Issue Editors

Konstantinos Kontis
School of Engineering,
University of Glasgow
UK

Spiros Pantelakis
Department of Mechanical Engineering
and Aeronautics, University of Patras
Greece

Editorial Office
MDPI
St. Alban-Anlage 66
4052 Basel, Switzerland

This is a reprint of articles from the Special Issue published online in the open access journal *Aerospace* (ISSN 2226-4310) from 2018 to 2019 (available at: https://www.mdpi.com/journal/aerospace/special_issues/8th_EASN-CEAS_Workshop)

For citation purposes, cite each article independently as indicated on the article page online and as indicated below:

LastName, A.A.; LastName, B.B.; LastName, C.C. Article Title. *Journal Name* **Year**, *Article Number*, Page Range.

ISBN 978-3-03921-485-3 (Pbk)
ISBN 978-3-03921-486-0 (PDF)

Contents

About the Special Issue Editors

Professor Kontis is a world-leading authority in aerodynamics, wind tunnel testing, flow control, shock physics, and diagnostics. He has published 108 journal articles, 1 book on shock waves (2 volumes, published by Springer), 6 book chapters, and 176 conference papers focusing on both fundamental scientific research and practical engineering applications. He is regularly invited to publish his conference papers in Special Issues of prestigious journals. He has received continuous EPSRC funding since 2001; other sources of funding include Royal Society, EU, Nuffield Foundation, USAF, USNavy, JSPS, and the industry. He has raised over £13.5M in external funding in the last 10 years, including over 60 separate collaborative projects with industrial partners—a testament to the industrial applicability of his work. The new developed methodologies are now being adopted by industry and universities worldwide. He has given 57 invited presentations and keynote lectures to peer-reviewed conferences and international advanced schools in the USA, Europe, India, China, and Japan.

Professor Pantelakis is Director of the Laboratory of Technology and Strength of Materials of the Mechanical Engineering and Aeronautics Department at the University of Patras. He served as Chairman of the Department of Mechanical Engineering and Aeronautics from September 2009 to September 2013. In September 2007 to August 2010, he was an Executive Board member of the Research Committee of the University of Patras. Furthermore, between September 2010 and August 2013, he was Vice-Chairman of the Executive Board of the Research Committee of the University of Patras. He is one of the founding members and, since 2008, Chairman of the European Aeronautics Science Network (EASN) Association. From 2006 to 2011, he was Representative of the European Aeronautics Academia in the plenary of ACARE, and from 2005 to 2011 he was Chairman of ACARE's Working Group on Human Resources and Chairman of the Board of Directors of the Greek Metallurgical Society from 2008 to 2011. Since 2015, he has been a member of the Board of Directors of the Hellenic Aerospace Industry. He has over 35 years' experience in the field of aerostructures and aeronautics materials during which he has been involved in over 100 international aeronautics research projects—in many of them, as Scientific Coordinator. He is a member of the Editorial Board and Guest Editor-in-Chief for a number of international scientific journals and Reviewer for many international scientific journals. Furthermore, he has been author or served as Editor in numerous books published by various international publishing houses. He is the author of more than 250 scientific publications in international peer-reviewed journals and conference proceedings. He has been Chairman of some dozens of international conferences and workshops, and founder of the Conference Series ICEAF (International Conference of Engineering Against Failure).

Editorial

Special Issue "8th EASN–CEAS Workshop on Manufacturing for Growth and Innovation"

Spiros Pantelakis [1,*] **and Konstantinos Kontis** [2,*]

1   Department of Mechanical Engineering and Aeronautics, University of Patras, Panepistimioupolis Rion, 26500 Patras, Greece
2   School of Engineering, University of Glasgow, James Watt Building South, University Avenue, Glasgow G12 8QQ, Scotland, UK
*   Correspondence: pantelak@mech.upatras.gr (S.P.); kostas.kontis@glasgow.ac.uk (K.K.)

Received: 29 July 2019; Accepted: 30 July 2019; Published: 30 July 2019

This Special Issue contains selected papers from works presented at the 8th EASN–CEAS (European Aeronautics Science Network–Council of European Aerospace Societies) Workshop on Manufacturing for Growth and Innovation, which was held in Glasgow, UK, 4–7 September 2018. About 150 participants contributed to a high-level scientific gathering providing some of the latest research results on the topic, as well as some of the latest relevant technological advancements.

Nine interesting articles, which cover a wide range of topics including characterization, analysis and design, as well as numerical simulation, are contained in this Special Issue. Trompeta et al. [1] assessed the critical multifunctionality threshold in order to optimize the electrical, thermal and nanomechanical properties of composites enhanced with carbon nanotubes for aerospace applications. The results showed an increased electrical conductivity up to seven orders of magnitude at a critical concentration. An increasing thermal stability was found as the multiwalled carbon nanotubes' (MWCNTs') concentration was increasing. Improved nanomechanical properties were exhibited. However, increasing further the MWCNTs' concentration led to decreased nanomechanical properties. In the paper of Donateo et al. [2], a methodology to quantify the synergies between power system architecture, energy and the actual battery power was proposed. Two cases were investigated: the case of a fixed-wing-tail-sitter Vertical Take-Off and Landing Unmanned Aerial Vehicle (VTOL-UAV), and a Medium-Altitude Long-Endurance Unmanned Aerial Vehicle (MALE-UAV). The results showed large improvements in the first case regarding the target performances. Sahin and Yaman [3] studied the synthesis, the analysis and the design of a special type of deployable mechanism, which is the scissor structural mechanism (SSM), for the morphing of the trailing edge of an aircraft wing. The analytical results showed a similar aerodynamic behavior with the desired NACA airfoils. Bergman et al. [4] developed an innovative scaled model for unmanned aircraft systems (UAS), named e-Genius-Mod. This model will facilitate the testing and assessment of new aviation technologies in a scaled version with an opportunity for free-flight demonstrations in relevant environments. Furthermore, the e-Genius-Mod system may close the gap between upstream research and flight demonstration, aiming at reaching a higher technical readiness level. In the study of Vrchota et al. [5], the boundary conditions applied to active flow circulation control were verified. The unsteady periodic flow inside the Suction and Oscillatory Blowing (SaOB) actuator was analyzed, using two Computational Fluid Dynamics (CFD) methods of ranging complexity (Unsteady Reynolds Average Navier–Stokes (URANS) and hybrid Reynolds Average Navier–Stokes (RANS)-Large Eddy Simulation (LES)). The simulation was experimentally validated. Although the obtained results are in a very good agreement with the experimental data, some differences still exist. Müller et al. [6] developed a novel numerical model that integrates the interaction between fluid and structure for modeling emergency landings on water, also referred to as ditching, in full scale. Specifically, this work focuses on the development of a suitable fluid-structure coupling framework and the generation of the structural model of the aircraft. Full-scale structural models for ditching are complex and need to account for damage. This results in severe nonlinearities in the model. Katsiropoulos et al. [7] have assessed the carbon footprint and the financial viability of different materials and manufacturing scenarios, as well as recycling scenarios, associated with the production of aeronautical structural components. For this purpose, the authors developed Life Cycle Analysis (LCA) and Life Cycle Costing (LCC) models and implemented them for the case of a helicopter's canopy production. The results showed that the optimal route both in terms of environmental and financial efficiency was the production of the canopy by using carbon fiber reinforced thermosetting composites and involving resin transfer molding (RTM) as the manufacturing process. Torres et al. [8] described the technological and scientific efforts on designing, manufacturing and testing validation for high-performance low-cost composite structures for Light Sport Aircrafts (LSA). The manufacturing and the assembly of a Light Sport Aircraft were developed by using Computational Fluid Dynamics (CFD), Finite Element Analysis (FEA) and Liquid Composite Manufacturing (LCM). A 15% difference in the performance of the aileron beam as compared to the simulation results was found, which was attributed to manufacturing errors. In the study of Dubovikov et al. [9], structural and manufacturing schemes for low-curvature pressurized fuselage panels were proposed, facilitating the possibility to provide high-weight efficiency for the airframes of prospective civil blended wing-body (BWB) aircraft. The ability to manufacture double-lattice panels was investigated and a relative model was developed. The results of the weight efficiency analysis of the double-lattice flat panels of a hypothetical BWB aircraft were presented. The results showed significant weight saving by the use of double-lattice panels as compared to the conventional composite analogue.

The editors of this Special Issue would like to thank the authors for their high-quality contributions and for making this Special Issue manageable. Additionally, the editors would like to thank Ms. Linghua Ding and the Aerospace editorial team.

Conflicts of Interest: The authors declare no conflict of interest.

References

1. Trompeta, A.-F.A.; Koumoulos, E.P.; Stavropoulos, S.G.; Velmachos, T.G.; Psarras, G.C.; Charitidis, C.A. Assessing the Critical Multifunctionality Threshold for Optimal Electrical, Thermal, and Nanomechanical Properties of Carbon Nanotubes/Epoxy Nanocomposites for Aerospace Applications. *Aerospace* **2019**, *6*, 7. [CrossRef]
2. Donateo, T.; De Pascalis, C.L.; Ficarella, A. Synergy Effects in Electric and Hybrid Electric Aircraft. *Aerospace* **2019**, *6*, 32. [CrossRef]
3. Şahin, H.L.; Yaman, Y. Synthesis, Analysis, and Design of a Novel Mechanism for the Trailing Edge of a Morphing Wing. *Aerospace* **2018**, *5*, 127. [CrossRef]
4. Bergmann, D.P.; Denzel, J.; Baden, A.; Kugler, L.; Strohmayer, A. Innovative Scaled Test Platform e-Genius-Mod—Scaling Methods and Systems Design. *Aerospace* **2019**, *6*, 20. [CrossRef]
5. Vrchota, P.; Prachař, A.; Hospodář, P. Verification of Boundary Conditions Applied to Active Flow Circulation Control. *Aerospace* **2019**, *6*, 34. [CrossRef]
6. Müller, M.; Woidt, M.; Haupt, M.; Horst, P. Challenges of Fully-Coupled High-Fidelity Ditching Simulations. *Aerospace* **2019**, *6*, 10. [CrossRef]
7. Katsiropoulos, C.V.; Loukopoulos, A.; Pantelakis, S.G. Comparative Environmental and Cost Analysis of Alternative Production Scenarios Associated with a Helicopter's Canopy. *Aerospace* **2019**, *6*, 3. [CrossRef]
8. Torres, M.; Piedra, S.; Ledesma, S.A.; Escalante-Velázquez, C.; Angelucci, G. Manufacturing Process of High Performance–Low Cost Composite Structures for Light Sport Aircrafts. *Aerospace* **2019**, *6*, 11. [CrossRef]
9. Dubovikov, E.; Fomin, D.; Guseva, N.; Kondakov, I.; Kruychkov, E.; Mareskin, I.; Shanygin, A. Manufacturing Aspects of Creating Low-Curvature Panels for Prospective Civil Aircraft. *Aerospace* **2019**, *6*, 18. [CrossRef]

Article

Assessing the Critical Multifunctionality Threshold for Optimal Electrical, Thermal, and Nanomechanical Properties of Carbon Nanotubes/Epoxy Nanocomposites for Aerospace Applications

Aikaterini-Flora A. Trompeta [1],[†], Elias P. Koumoulos [1],[†], Sotirios G. Stavropoulos [2],[†], Theodoros G. Velmachos [2],[†], Georgios C. Psarras [2],[†] and Costas A. Charitidis [1],[*],[†]

1   Research Unit of Advanced, Composite, Nano-Materials & Nanotechnology, School of Chemical Engineering, National Technical University of Athens, Zographos, 15780 Athens, Greece; ktrompeta@chemeng.ntua.gr (A.-F.A.T.); elikoum@chemeng.ntua.gr (E.P.K.)
2   Smart Materials & Nanodielectrics Laboratory, Department of Materials Science, University of Patras, 26504 Patras, Greece; Stavropoulos@upatras.gr (S.G.S.); teovelmachos@gmail.com (T.G.V.); psarras@upatras.gr (G.C.P.)
*   Correspondence: charitidis@chemeng.ntua.gr; Tel.: +30-210-772-4046
†   These authors contributed equally to this work.

Received: 17 December 2018; Accepted: 7 January 2019; Published: 10 January 2019

Abstract: Epoxy composites are widely used in primary aerospace structures, where high impact damage properties are necessary. However, challenges appear when multiple functionalities, including electrical and thermal conductivity, are needed in parallel with increased mechanical properties. The current study aims at the assessment of a critical concentration of multiwalled carbon nanotubes (MWCNTs), incorporated in epoxy resin, which will indicate a threshold for optimal electrical, thermal and mechanical properties. For the evaluation of this optimal concentration, electrical conductivity, thermal stability and nanomechanical properties (Young modulus and nanohardness) have been assessed, for epoxy nanocomposites with 0 to 15 parts per hundred resin per weight (phr) MWCNTs. Percolation theory was applied to study the electrical conductivity for different contents of MWCNTs in the epoxy nanocomposite system. Thermogravimetric analysis was employed for the assessment of the epoxy composites' thermal properties. Nanohardness and elastic modulus were measured, and the hardness versus modulus index was calculated. Emphasis was given to the dispersion of MWCNTs in the epoxy matrix, which was assessed by both microscopy techniques and X-ray micro–computed tomography. A correlation between the optimum dispersion and MWCNTs content in terms of electrical conductivity, thermal stability, and nanomechanical properties revealed a threshold concentration at 3 phr, allowing the manufacturing of aerospace structures with multifunctional properties.

Keywords: carbon nanotubes; electrical properties; multifunctionality; nanomaterials; nanomechanical properties; polymer nanocomposites; thermal stability; threshold concentration

---

1. Introduction

Nanocomposite materials combining a polymer with nanofillers have attracted research interest the last decades, due to the combined advantages they offer [1]. Carbon nanotubes (CNTs) have been widely used as nanofillers in epoxy composites, since they provide improved electrical, thermal, and mechanical properties in the final product [2]. The extraordinary properties of CNTs offer the ability to manufacture conducting polymers at low concentrations of the nanofiller, without altering the performance of the polymer, in comparison with the use of other additives such as carbon black, which requires higher concentrations for an effective functionality [3]. Indeed, if the desired performance can be achieved at low concentrations, novel lightweight materials can be fabricated, to be used in a range of advanced applications. The benefits of composite materials are clearly depicted on aerospace applications which require lightweight, high-strength, high-stiffness, and highly fatigue-resistant materials [4]. The use of composites within the aerospace industry has revolutionized aircraft design, and has already resulted in significant fuel savings across many fleets (up to 20%) compared with aluminum. Composites have been used in aircraft parts, as presented in Table 1.

Table 1. Structural parts of aircrafts where composite materials can be used, according to Ghori et al., 2018 [5].

Primary Structural Parts	Secondary Structural Parts
Wings	Doors
Tails surfaces	Antenna dishes
Fuselages sections	Center wing boxes
Rocket motor castings	Landing gear doors
Engine nacelles and cowls	Floor beams
Horizontal and vertical stabilisers	Tall cones
Pressure bulkheads	Flap track panels

According to the above, the aerospace industry can exploit the extraordinary properties of CNTs to enhance the mechanical properties needed for aerospace constructions. Researchers have already studied the effect of CNTs on the mechanical properties of reinforced epoxy resin composites. Taraghi et al. studied the impact response for reinforced composites containing CNTs, and found that the impact strength can increased up to 45% [6]. Several studies have also investigated the mechanical properties of such composites, focusing on the elasticity [7], tribology [8], toughness [9], and damage [10] of the specimens. Nanomechanical measurements through nanoindentation have also been used for the assessment of the mechanical properties of nanocomposites based on epoxy resin which have been reinforced with singlewalled CNTs (SWCNTs) [11], since nanomechanics are of great importance in the aeronautics and aerospace industry [12].

The aerospace industry can also benefit from the exploitation of cost-effective nanocomposite materials with tailored conductivity [13]. Carbon nanotubes have been distinguished from other nanomaterials, due to their extraordinary electrical and thermal properties. The study of Alsafee et al. [14] demonstrated the intrinsic potential of multiwalled carbon nanotubes (MWCNTs) to change the electrical behavior of an epoxy resin. It was revealed that a small quantity of MWCNTs can considerably modify the electrical properties, and transform the epoxy matrix from an insulator to a semiconductor. This finding can be exploited by the aerospace industry as a strategy for lightning strike protection [15]. Regarding the thermal properties, there is still lot of space to investigate the connection between glass transition temperature and mechanical properties [16]. The aerospace industry is continuously working to develop lightweight and strong materials with increased thermal stability that can stand high temperatures [17].

Despite the fact that the above properties have individually been investigated extensively, combined studies aiming to assess a threshold concentration for the achievement of a transition in electrical and thermal properties together with an improvement in the nanomechanical properties for epoxy-based nanocomposites are really scarce. The work of Allaoui et al. [18] was an attempt to investigate electrical and mechanical properties, by means of tensile tests and AC/DC spectroscopy measurements. However, the mechanical testing had not been executed in the nanoscale to detect phenomena that are occurrent in such dimensions. Thus, this study, employing nanomechanical testing, thermal degradation evaluation, and electrical conductivity measurements, aims to meet the aforementioned properties and conclude on an optimum reinforcement concentration for the epoxy nanocomposite material, to be used in aerospace applications, as a nanomodified resin system of the preparation of reinforced epoxy composites. In order to eliminate the chemical effects [11,19] and test the performance of the pristine MWCNTs, unmodified MWCNTs synthesized in-house have been chosen as nanofillers in this study. By this, the performance of existing masterbatches of epoxy resin with pristine CNTs that are commercially available and are potentially being used in aerospace applications will be assessed.

2. Materials and Methods

2.1. Materials

The chosen method for the synthesis of MWCNTs used in the preparation of the composites was thermal chemical vapor deposition (T-CVD). The supported catalyst approach was employed, as first reported in Trompeta et al.'s study [20], which offers higher productivity and yield, resulting in lower production costs (estimated at less than 1 euro/g). For the growth of MWCNTs, a catalyst containing Fe particles at 20 wt. % was synthesized in-house with the precipitation method, using a mix of zeolite and Al_2O_3 (50:50) as an absorbent substrate. Silicon (Si) wafers of N type, P doped, with a crystallographic orientation of (100) were used as substrate to deposit the catalyst prior its insertion in the reactor. The catalyst was placed on the polished side of the Si wafers, which were cut on the appropriate dimensions to fit inside the reactor, i.e., 3.4×10 cm. For each experiment, 0.3 to 0.8 g catalyst was used. Prior to the initiation of the experiment, purging with N_2 was necessary for 30 min to remove the oxygen from the reactor, and preheating at 550 °C took place for 1 h, for the calcination of the catalyst. Afterwards, the CVD reactor was heated up to 700 °C. Then, the reaction was started while feeding the system with acetylene with a volumetric flow of 100 mL/min. Simultaneously, the nitrogen flush was continued with a volumetric flow of 230 mL/min. The reaction for each batch lasted from 1 to 8 h. The product was received in a sponge-like form which was easily manipulated and turned into powder through mechanical grinding, to be used for the preparation of the nanocomposites. Each batch resulted in a production of ~6.5 g, which led to the repetition of the process for five times in order to obtain the necessary MWCNT quantity for the composites. Full repeatability was evident between the experiments, as can be seen in Figure 1, which depicts the morphology of the synthesized MWCNTs as received from two different batches. As it can be seen, both samples consisted of bundles of MWCNTs which varied in diameter from ~45 to 95 nm and had lengths of up to 5 μm, which classifies them as "short" MWCNTs [21].

Of all resins used in the aerospace industry, epoxy resins have gained the widest acceptance [22] since they offer versatility that is unattainable among other polymers, such as phenolic resins, polyurethanes, high-performance polymers (polyarylene ether nitrile (PEN) [23], addition-type polyimides (PI), i.e., polyamide-imide (PAI), polyphthalamide (PPA), polyethylenimine (PEI), polysulfones (PSU, PESU, PPSU), polyether ether ketone (PEEK), polyphenylene sulfide (PPS), polyvinylidene fluoride (PVDF), perfluoroalkoxy alkane (PFA), and liquid crystal polymers (LCP) [24]). Epoxy resins have a wide range of viscosities. Thus, they are adaptable to a wide variety of manufacturing processes. For example, filament winding and pultrusion using low-viscosity resins, or dry lay-up processing with solid-form resins [22]. The epoxy polymer matrix was prepared by mixing the low-viscosity (500–700 MPa·s) epoxy resin (Araldite® GY 257, based on Bisphenol A modified with an aromatic glycidyl ether), one of the most commonly used epoxy resins in aerospace applications [22], with hardener diethylenetriamine in a ratio of 1:0.05. This low-viscosity resin was chosen in order to avoid difficulties on the processability of the nanocomposites on high MWCNT content.

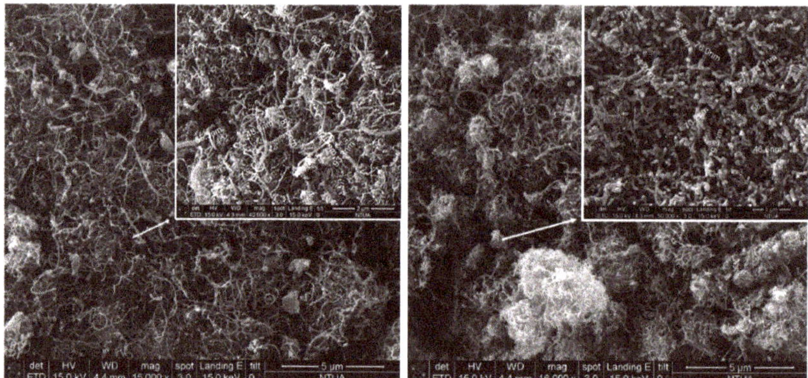

Figure 1. Entangled multiwalled carbon nanotubes (MWCNTs) with diameters ranging between 45 and 95 nm, used as nanofillers in the current study.

2.2. Composites Preparation

To prepare the composites, the entangled MWCNTs were first grinded through mechanical milling. Subsequently, the chosen quantity of MWCNT powder was added to the epoxy system, which included both the first part of the liquid epoxy resin and the hardener in a volume of 50 mL, in order to prepare admixtures with different concentrations of parts per hundred resin per weight (phr): 0.1 phr (sample EP01), 0.5 phr (sample EP05), 1 phr (sample EP1), 2 phr (sample EP2), 3 phr (sample EP3), 5 phr (sample EP5), 10 phr (sample EP10), and 15 phr (sample EP15). Higher loadings of up to 80 wt. % have been achieved with the procedure developed in the work of Schilde et al. [25]. However, the described procedure was out of the scope of the current study. The admixture was prepared via high-shear mechanical mixing, using a sawtooth impeller of 2 cm diameter, in 1000 rpm for 30 min for each sample, including the preparation of a bare resin sample reference for comparison reasons. According to the literature [26], high-shear mixing has proven to be more effective in comparison with other methods [27], such as ultrasonication, for the preparation of nanocomposites, since it enables the dispersion and intercalation of the nanofillers in the polymer matrix due to the temperature rise [28,29]. Despite the fact that this technique is not new, it can be considered cost- and time-effective, and it simulates the industrial production of polymer masterbatches, used in the polymer, paint, coatings, and composites industries [30]. The mixtures of each case were poured into small molds of 3 cm diameter, in duplicates, and were placed under vacuum overnight in order to absolve possible bubbles, since the first component of the resin system is a solution in a mono-functional reactive diluent. Afterwards, the composites were cured in a furnace at 70 °C for 4 days. The surface of the first replicate of each of the obtained specimens was grinded in order to be flattened, and one of the two sides was polished for the nanomechanical testing. For the electrical conductivity measurements, the second replicate was grinded from both sides, in order to obtain a disc-like specimen of about 2 mm thickness.

2.3. Characterization Techniques

2.3.1. Optical Microscopy

Optical microscopy was executed using an Axio Imager A.2m (Carl Zeiss MicroImaging GmbH, Jena, Germany) in the transmitted light configuration, using the bright field filter, in two magnifications (×100 and ×500). Multiple images were acquired and combined (z-stitching) for high depth of field (DOF) images, using Zerene Stacker software (free trial version, Zerene Systems LLC, Richland, WA, USA).

2.3.2. Scanning Electron Microscopy

The morphologies of the MWCNTs used in this study, as well as the cryo-fractured surfaces of the nanocomposites (via liquid nitrogen), were examined by scanning electron microscopy (SEM) using a Nova^{TM} NanoSEM 230 (FEI company, Hillsboro, OR, USA) microscope with a W (tungsten) filament, with all images acquired in secondary electron mode using an operating voltage of 15 kilovolts (kV).

2.3.3. X-ray Microcomputed Tomography (Micro-CT)

Micro-CT examination has been proven to be ideal for the study of reinforced composites, especially when their internal structures need to be assessed for demanding applications, such as aeronautics. Thus, the structures of the specimens were investigated by a compact desktop micro-CT, 3D X-ray scan system, SkyScan 1272 (Bruker, Kontich, Belgium). The system consisted of a microfocus sealed X-ray source L11871-02 (Hamamatsu Photonics K.K, Hamamatsu, Japan) which operated at 20–100 kV and 10 W (< 5 μm spot size @ 4 W). The X-ray detector, which had a resolution of 11 Mp (4032 × 2688 pixels) was coupled to the scintillator. The distance between the rotation axis and the radiation source was set from 6 to 12 mm, and the pixel resolution was set to 2016 × 1344 in order to obtain reliable images. The X-ray tube voltage was set to 40 kV and the X-ray tube current was set to 250 μA for all the specimens. After the acquisition of the X-ray projection images, the 3D image was reconstructed by the image processing software of the X-ray CT system (InstaRecon®, Bruker, Kontich, Belgium).

2.3.4. Electrical Measurements

Electrical measurements were conducted by means of Broadband Dielectric Spectroscopy (BDS) in a wide frequency (10^{-1}–10^{6} Hz) and temperature range (30–150 °C), via an Alpha-N Analyzer, high resolution dielectric analyzer (Novocontrol Technologies & Co. KG, Montabaur, Germany). The applied voltage amplitude was 1 V in all cases. Examined specimens were put, in a parallel plate form, in the BDS 1200 dielectric cell and subjected to isothermal scans with a temperature step of 10 °C. Temperature was controlled via a Novotherm system, and real-time data acquisition was performed by employing Windeta software (Novocontrol Technologies & Co. KG, Montabaur, Germany). The software, dielectric cell, and temperature control system were also supplied by Novocontrol Technologies.

2.3.5. Thermogravimetric Analysis

Thermal properties were examined using a STA 449F5 Jupiter ®thermogravimetric analyzer (Netzsch, Selb, Germany) in a temperature range from 25 to 1000 °C, under a constant air flow of 50 mL/min, while heating rate was set to 10 °C per minute.

2.3.6. Nanomechanical Assessment

Through their responses to local loading, the nanomechanical integrities of the materials were assessed through a Hysitron (Minneapolis, MN, USA) TriboLab®Nanomechanical Test Instrument, which recorded the displacement with a load resolution of 1 nN and displacement resolution of 0.04 nm. In all presented data, a total of 10 indents were performed so as to determine the average hardness (*H*) and elastic modulus (*E*) values, with spacing of 50 μm, at room temperature with 45% humidity. The closed-loop feedback control option was selected. All measurements were performed using the standard three-sided pyramidal Berkovich diamond tip indenter (radius of 100 nm) [31,32].

From the half-space elastic deformation theory, H and E values were extracted from the experimental data (load–displacement curves) using the Oliver-Pharr (O&P) model [33]. The expressions that calculate E, extracted from load–displacement curves, are based on Sneddon's [34] elastic contact theory:

$$E_r = \frac{S\sqrt{\pi}}{2\beta\sqrt{A_c}} \tag{1}$$

where S is the unloading stiffness (initial slope of the unloading load–displacement curve at the maximum displacement of penetration (or peak load)), Ac is the projected contact area between the tip and the substrate, and β is a constant that depends on the geometry of the indenter (β = 1.167 for the Berkovich tip [33,34]). Nanoindentation H refers to the mean contact pressure required. This hardness, which is the contact hardness (Hc), is actually dependent upon the area geometry of the tip indenter, as described in Equations (2)–(4):

$$H_c = F/A \tag{2}$$

where

$$A(h_c) = 24,5h_c{}^2 + a_1 h_c + a_{1/2}h_c{}^{1/2} + \ldots + a_{1/16}h_c{}^{1/16} \tag{3}$$

and

$$h_c = h_m - \varepsilon\frac{P_m}{S_m} \tag{4}$$

where h_m is the total penetration displacement of the indenter at peak load, P_m is the peak load at the indenter displacement h_m, and ε is an indenter geometry constant, equal to 0.75 for the Berkovich indenter [32–35]. Prior to indentation, the area function of the indenter tip was calibrated in a fused quartz/silica (standard material) [36].

3. Results and Discussion

3.1. Dispersion Characterization

In order to fully exploit the properties of CNTs and to transfer the mechanical, electrical, and thermal stability enhancements to the polymer matrix, an effective dispersion should be achieved [37]. A strong interfacial bonding between the inorganic phase of the CNTs and the organic phase of the polymer is the main challenge for the formation of successful polymer nanocomposites. To achieve this prerequisite, a homogenous dispersion of CNTs in the matrix is also important [38]. However, the entanglement of CNTs leads to agglomeration in the micrometer scale [39]. The following results, retrieved from advanced microscopy techniques and micro-CT, prove the effectiveness of the CNT dispersion in the epoxy matrix [40].

3.1.1. Optical Microscopy

For the evaluation of the dispersion of MWCNTs in the epoxy resin, optical microscopy was chosen for an initial assessment [19]. From the prepared samples, the transparent one was chosen (EP01: 0.1 phr MWCNTs) for observation in the transmittance mode. The results are depicted in Figure 2, in two different magnifications ($\times$100 and $\times$500, respectively). As it can be seen, MWCNT bundles were homogenously dispersed within the epoxy matrix. Agglomerates of about 30–50 µm were observed, and they can improve the impact resistivity of the composites since they can absorb the energy from the stress.

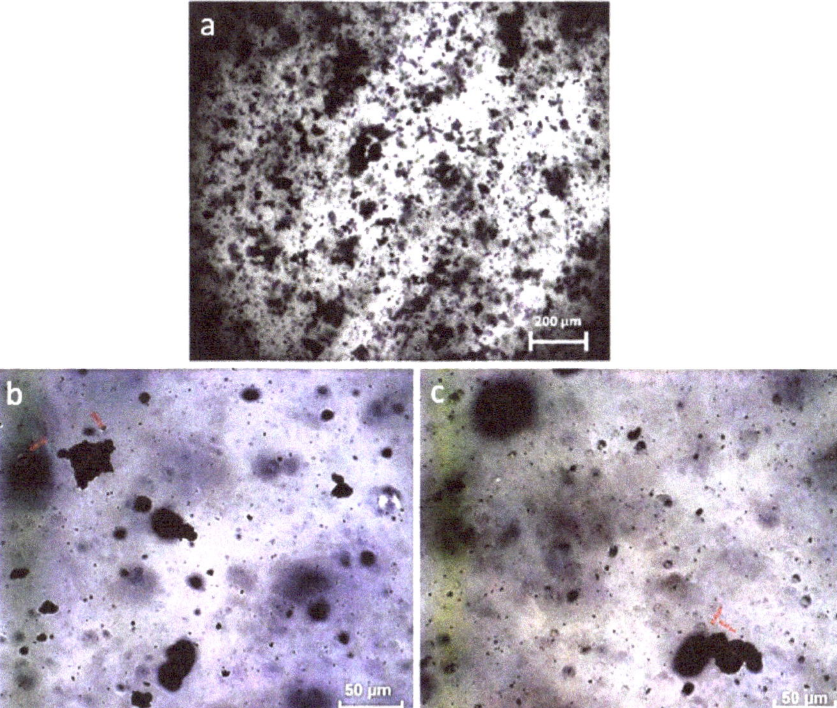

Figure 2. (**a**) Optical microscopy image (transmittance mode) for EP01 where the dispersion of MWCNTs is observed; magnification: ×100); (**b** and **c**) Optical microscopy image (transmittance mode) for EP01 at ×500 magnification. The dimensions of the MWCNT bundles are up to ~30 µm.

3.1.2. Study of Cryo-Fractured Surface

The cryo-fractured surfaces of three indicative samples with a low (0.5 phr), medium (3 phr), and high (15 phr) concentration were studied under SEM, and the results are shown in Table 2. In the low magnification (×600), it was possible to have an estimation of the dispersion of MWCNTs inside the composite. It is clear that the low concentration does not enable the existence of MWCNTs in every part of the specimen. On the other hand, when the concentration increased, MWCNTs were uniformly dispersed in the whole matrix. Also, the fracture lines were evident in the low magnifications. It could be observed that the fracture lines are blocked around the MWCNT areas, as shown for the EP05 and EP3 samples. In this case, MWCNT agglomerates work as barriers for the crack propagation. In higher magnifications (×10,000, ×20,000), the structure of MWCNT agglomerates and the alignment of the MWCNTs in the polymer matrix could be observed. Finally, the magnification of ×40,000 enabled the observation of MWCNTs as well as their dimensioning. It may be remarked that the observed diameters of MWCNTs inside the composite were slightly increased (MWCNTs seem to become thicker) in comparison with the measured diameters that are presented in Figure 1. This indicates that the polymer has been wrapped around MWCNTs with non-covalent bonds.

Table 2. SEM images of the cryo-fractured surfaces of the nanocomposites.

Mag.	Bare Resin	EP05	EP3	EP15
×600				
×10,000				
×20,000				
×40,000				

3.1.3. Evaluation of Dispersion through Micro-CT

According to Kim et al. [37], the actual dispersion of nanofillers within polymer composites cannot be easily evaluated using SEM. On the contrary, the real dispersion could be clearly observed in the 3D micro-CT images. Thus, in Table 3, the dispersion of MWCNTs in selected samples is depicted, after the micro-CT scanning. In each case, a cuboid representative volume that was considered to be representative of the entire volume was identified for detailed analysis. As it can be seen, for concentrations of up to 2 phr, the dispersion of MWCNTs in the epoxy matrix was too sparse. However, for concentrations of 3 phr and higher, a homogenous dispersion could be achieved, covering the entire volume of the specimen. Thus, a network can be formatted, as possible conjunctions are feasible. Moreover, micro-CT was implemented in order to assess the structural integrity of the samples and the detection of possibly remaining voids that occurred during the degassing process. After the examination of all samples, it could be remarked that voids were not detected, as they would have been marked with a different color (red) due to their density difference with the epoxy resin and the MWCNTs.

Table 3. Microcomputed Tomography (Micro-CT) images of MWCNT–epoxy samples.

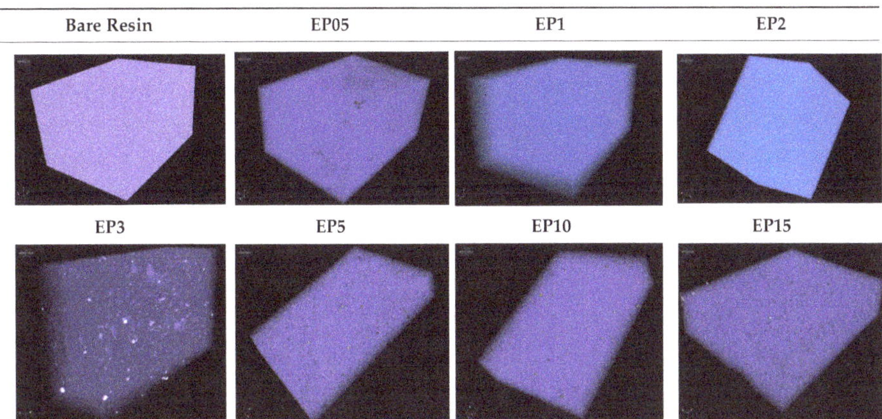

Bare Resin	EP05	EP1	EP2

EP3	EP5	EP10	EP15

3.2. Electrical Conductivity Evaluation

In Figure 3, the AC conductivity at two different temperatures is shown as a function of the applied frequency for nanocomposites and neat epoxy. AC conductivity was evaluated from the dielectric loss factor (ε'') via Equation (5):

$$\sigma_{AC} = \varepsilon_0 \omega \varepsilon'' \tag{5}$$

where ε_0 is the permittivity of free space and ω is the angular frequency of the applied field.

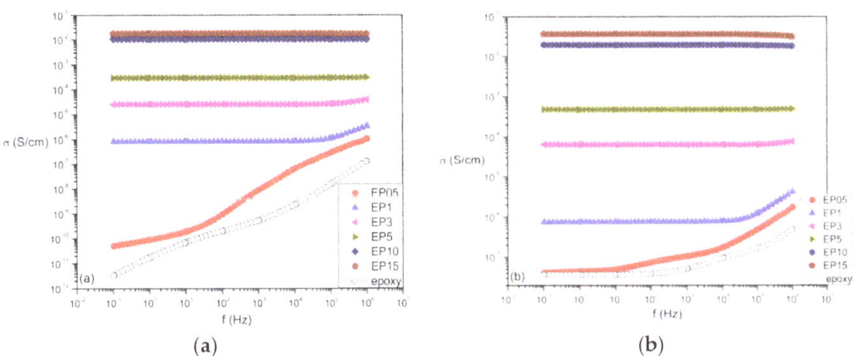

Figure 3. Variation of conductivity with frequency at: (**a**) 30 °C and (**b**) 90 °C.

Conductivity values increase with filler content, while being temperature- and frequency-dependent. In the study of Bellucci et al., the resistivity of epoxy resin with 0.5 wt. % CNTs with similar geometrical characteristics as in our study was tested, and a reduction 10^3 was found [38]. In our case, the neat epoxy and EP05 specimens exhibited a dispersion of conductivity with frequency. At low frequencies, σ_{AC} values tended to be constant, approaching their DC conductivity value, whereas after a characteristic frequency, conductivity followed an exponential dependence. In the low-frequency regime, the applied electric field keeps its direction for a longer time compared to the high-frequency regime, thus allowing charge carriers to migrate longer distances within the material. However, the insulating nature of the matrix exerts strong restrictions to this motion, diminishing the number of carriers which are able to overcome the existing potential barriers. Therefore, a limited number of charge carriers migrate at relatively longer distances. At the high-frequency region, the alternation of the field does not provide sufficient time for charge carriers to migrate at relatively long distances. On the other hand, charges are able to move or hop between adjacent conductive sites. The outcome of this process is increased values of conductivity, since a large number of charges participate by hopping between neighboring positions, covering very short distances [39,40]. At higher temperatures, the dispersion of conductivity with frequency decreases, because thermal agitation facilitates the migration of charge carriers, as shown in Figure 3b. This facilitation is also expressed via the enhanced values of conductivity.

Concerning the effect of MWCNT content, an abrupt increase of conductivity, of several orders of magnitude, was observed for the specimen EP3, with 3 phr MWCNT content, over the whole frequency and temperature range. This increase, in the case of 0.1 Hz and 30 °C, exceeded the seven orders of magnitude, providing a strong indication that the examined nanocomposites underwent a transition from insulating to conductive behavior. Moreover, the influence of frequency upon conductivity values vanished with increasing MWCNT content, and apparently became negligible for concentrations higher than 1 phr, implying that the critical concentration for the transition from insulating to conductive behavior lies between 1 and 3 phr filler content. This is also in agreement with [19]. The insulator-to-conductor transition is considered as a critical phenomenon, and can be described by means of percolation theory according to Equation (6):

$$\sigma \approx (P - P_C)^t \tag{6}$$

where P is the content of the conductive phase, P_C the critical concentration (also known as the percolation threshold), and t is a critical exponent related to the dimensionality of the conduction process [39,40]. At the critical concentration, a conductive path is formed, allowing charge carriers to migrate through the whole nanocomposite [41].

The variation of conductivity with MWCNT content and frequency at two temperatures is depicted in the plots of Figure 4.

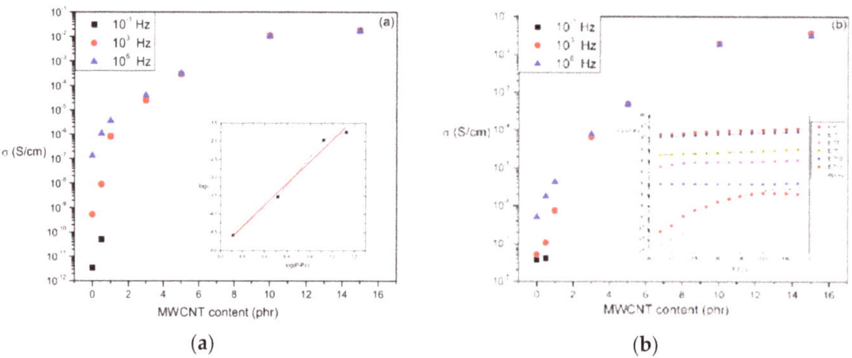

Figure 4. Variation of conductivity with MWCNT content and frequency at: (**a**) 30 °C (the inset provides an example of percolation threshold determination), and (**b**) 90 °C (the inset depicts the variation of conductivity with temperature at f = 0.1 Hz).

It can be observed that, within a narrow variation of filler content, a significant alteration of conductivity exists in all cases. The effect of temperature is more pronounced at low frequencies and below the critical concentration, since the low alternation of the field and the thermal agitation provide sufficient time and energy for charge carriers to migrate within the nanocomposite. Fitting the experimental data with Equation (6) in its double logarithmic form allowed for the determination of the critical concentration and exponent, which in the case of 0.1 Hz and 30 °C were 1.7 and 2.8, respectively. The determined value of the percolation threshold confirmed that the EP3 nanocomposite had undergone the insulating-to-conductive transition, while the value of the critical exponent appeared to exceed the value predicted by classical percolation theory, which for a three-dimensional conduction process is two. Deviations from the predicted value by classical percolation theory are often observed, and can be attributed to the differences between a perfect lattice with spherical conductive inclusions having no mutual interactions, and a real CNT-polymer system. The dynamic nature of the nanocomposite, along with interactions between the conductive MWCNTs and the composite's phases, contribute to this discrepancy [39–41].

As illustrated in Figure 3, the shapes of the recorded spectra of conductivity as a function of frequency, especially for the neat epoxy and EP05 systems, imply hopping conductivity as the dominating conductance mechanism [26,41]. The situation changes above the percolation threshold, where the effect of frequency upon conductivity values diminishes rapidly. At filler concentrations higher than the critical one, a conductive path (or paths) is created within the specimens, through which carriers can percolate the whole sample. Thus, another conduction mechanism occurs, and conduction is carried out via the geometrical contacts of MWCNTs in the path or paths. The inset of Figure 4b depicts the dependence of conductivity on temperature, at a constant frequency of 0.1 Hz, for the same set of specimens. Interestingly, the influence of temperature below and above the percolation threshold, as expressed by the differential $d\sigma/dT$, changed. Below the critical concentration, $d\sigma/dT$ is positive and conductivity values increased, indicating the dielectric/insulating nature of the composites. At these low filler contents, conductive paths are incomplete and charges are migrating via hopping conductivity. Above the critical concentration, $d\sigma/dT$ decreased approaching zero, since conduction via the created paths came into play. It should be noted that in the cases where metallic-type conduction takes place, the sign of $d\sigma/dT$ becomes negative. The absence of negative values of the differential $d\sigma/dT$ in the recorded data suggested that metallic-type conduction was not the dominating process, and hopping conductivity also appeared above the percolation threshold.

3.3. Thermal Decomposition Evaluation

Figure 5 shows TGA curves of the bare epoxy and MWCNT/epoxy composites from 380 to 520 °C in air atmosphere, which is the focus-of-interest range from the full TGA curves, where the thermal degradation phenomena take place. The thermal decomposition profile of the bare epoxy resin consisted of a drastic weight loss at around 360 °C, which corresponds to the degradation of bisphenol A groups [42]. In principle, MWCNTs showed good thermal stability as reinforcing agents, with a decomposition temperature of around 520 °C. They were completely decomposed at around 690 °C [43]. This good thermal stability is attributed to the high crystallinity of MWCNTs, due to their graphitic structure. This resulted in a higher thermal stability of the epoxy composite as the content of MWCNT increased, which is in accordance with the study of Si-Eun et al [44]. For a specific temperature, the weight loss of the composites with higher MWCNT phr was lower. Moreover, two distinct regions can be observed: The first one includes the lower concentrations of 0.1 to 2 phr content, while the second one refers to higher concentrations of 3 to 10 phr content. It may be remarked that the thermal stability increased three times in comparison to the bare resin with the 3 phr content, which is a threshold that comes in accordance with the electrical conductivity measurements. Finally, it should be mentioned that the difference observed in the inset graph at 200 °C, which corresponds to the bare resin (black squares), is a typical behavior of epoxy resins, which is attributed to moisture loss and entrapped solvent remaining in the sample, according to Baldissera et al. [45].

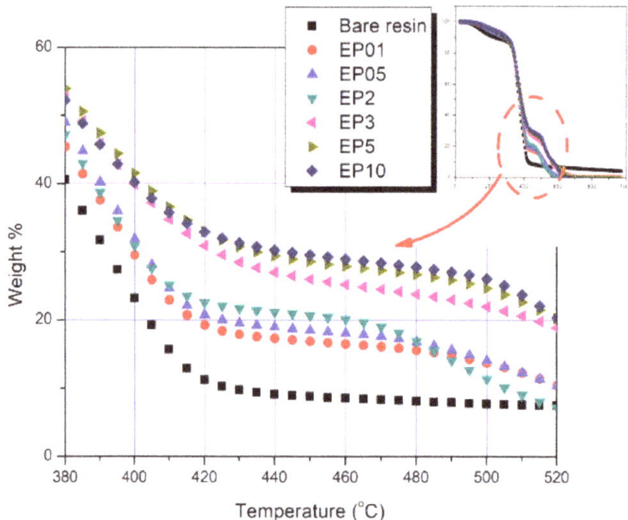

Figure 5. Thermogravimetric curves of nanocomposites.

3.4. Nanomechanical Characterization

Nanomechanical testing can overcome obstacles that are evident in the case of traditional mechanical measuring methods, such as tensile tests and microhardness testing. In parallel, the study of the local mechanical properties in a small volume can be achieved through nanoindentation, taking into account the material's microstructural gradients. The tiny contact area between the indenter and the sample's surface leads to high stress under specific loads. At low loads, a very prominent phenomenon is the so-called "Indentation Size Effect" (ISE), due to imperfection in tip geometry. This effect may cause deviations in the accurate measurement of hardness values [46].

Prior to nanoindentation and in order for the areas selected for being probed to be considered adequate (i.e., representative of the sample and of low roughness), the surfaces of the nanocomposites had been carefully examined (also for selection purposes) through Scanning Probe Microscopy (SPM) imaging (70 μm × 70 μm). Representative images, together with the roughness assessments, are presented in Figure 6.

Hardness and modulus values were found to deviate at surface regions (~0–400 nm), probably due to the roundness of the tip and ISE, tending to reach constant values as shown in Table 4.

The graded surface structures due to the existence of MWCNTs, as well as the MWCNT bundling, can be considered as possible contributors to the wide range of hardness and modulus values obtained from the nanoindentation measurements. In parallel, the adhesive forces between the tip and the sample also affected the nanoindentation results. Improved nanomechanical properties were revealed for the cases of all samples, except for samples EP5 and EP10 which exhibited decreasing H and E values, leading to significantly lower nanomechanical properties values, indicating deterioration. This is illustrated in Figure 7. When comparing with other concentrations, the case of EP3 was the optimal concentration threshold that strengthened the epoxy matrix.

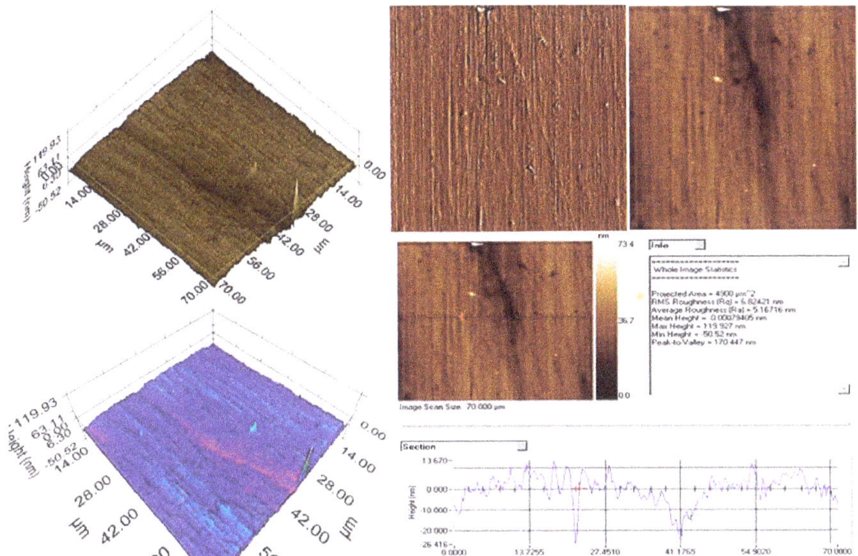

Figure 6. Representative SPM imaging of all samples, together with roughness assessments.

Table 4. Hardness and modulus bulk values of MWCNT-epoxy samples.

Sample	Hardness (H) (MPa)	Elastic Modulus (E) (GPa)
EP01	60 ± 3	0.8 ± 0.4
EP05	80 ± 3	2.0 ± 0.1
EP1	80 ± 4	2.0 ± 0.2
EP2	90 ± 5	2.0 ± 0.1
EP3	140 ± 6	2.4 ± 0.1
EP5	50 ± 4	0.7 ± 0.1
EP10	40 ± 3	0.5 ± 0.1
EP15	140 ± 5	2.0 ± 0.1

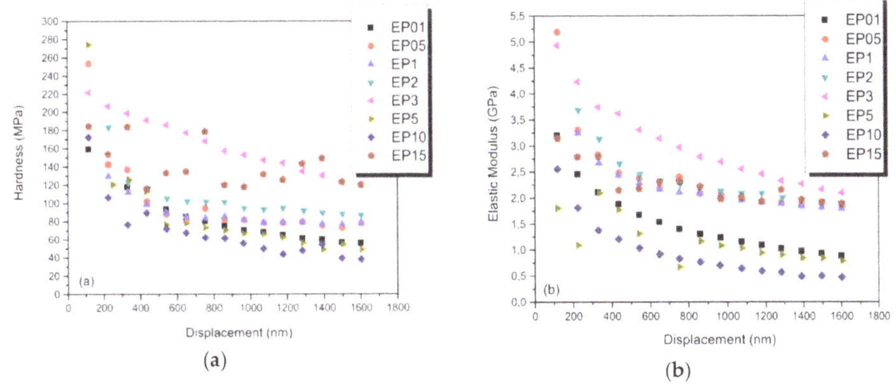

(a) **(b)**

Figure 7. Nanomechanical properties, namely (**a**) hardness and (**b**) modulus, of nanocomposite samples.

As it can be seen, a low concentration of MWCNTs in the epoxy matrix provides a satisfying interfacial interaction that leads to the stretching of the epoxy molecules which have been attached to the MWCNTs, as also proven by the SEM cross-sectional analysis of the cryo-fractured surfaces. Additionally, a tube–tube slip can occur in the individual MWCNT rich-phase clusters. As MWCNT concentration was increased (EP01 to EP3 sample), there was a large increase in the stiffness of the system, which resulted in a reduction in the energy dissipated during the experiment. However, there were more tube–tube slips in the system at higher concentrations, since percolation, coupled to a better dispersion, leads to a higher contact surface among MWCNTs. For the EP5, EP10, and EP15 samples, dissipated energy increased—increased elasticity is also evidenced in Figure 8a, but with increased deviation relating with dispersion.

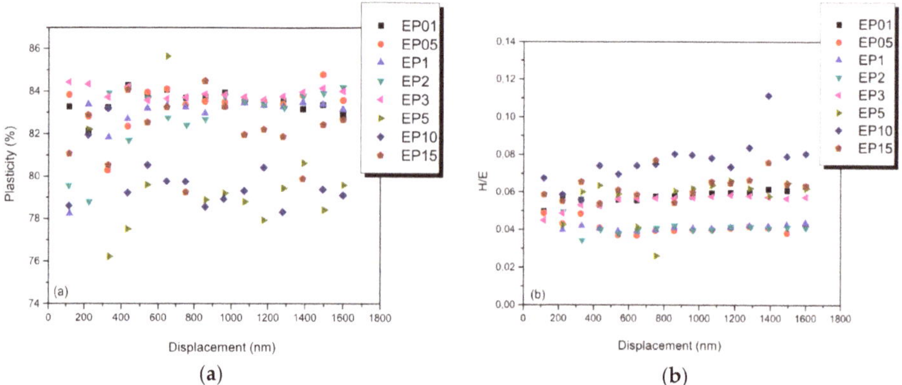

Figure 8. (a) Plasticity index obtained from load–unload curve area calculation, and (b) hardness-to-modulus ratios as indexes of resistance to wear.

It has been observed that the mechanical properties tend to deteriorate with increasing MWCNT content, after the threshold concentration of 3 phr. This fact is attributed to an extensive tube–tube slip mechanism, while nanotube contacts are unavoidable because of the sticky inter-particle tendency. Thus, these contacts (acting as knots) give rise to micron-sized agglomerates acting as mechanical defects for the resulting composite [47–49].

To evaluate the tribological properties of the nanocomposites, the hardness/elastic modulus ratio (H/E) has been calculated. The high values of the ratio are expected at higher stresses (usually evident in hard materials), whereas a low H/E indicates lower stresses which are distributed in the cross-section of the composite (found in softer materials) [50]. When H/E is high, a good wear resistance can be achieved [51,52], which has also been proven for polymeric materials. In Figure 8b, a change in the H/E slope is evident. This deviation reveals the strengthening of the nanocomposite with the addition of a critical amount MWCNTs. On the other hand, lower deviation of H/E values can prove the good dispersion of the MWCNTs in the polymer, with an almost-stable behavior at resistance to wear.

4. Conclusions

The potential use of MWCNT/epoxy nanocomposites in aerospace applications was investigated in this study, taking into account the main properties that are important for the materials choice in such constructions. The intrinsic potential of MWCNTs to considerably modify the mechanical and electrical behaviors of a polymer matrix, in low contents, was evident. MWCNTs in a content of 3 phr managed to improve the performance of epoxy composites, in terms of mechanical properties, thermal stability, and electrical conductivity, without the need of any further chemical process. The main findings summarized below are consistent with the formation of a MWCNT network over the percolation threshold, leading to MWCNT slipping:

- From the optical microscopy in transmittance mode, it could be observed that MWCNT bundles were homogenously dispersed within the epoxy matrix for the samples with low MWCNT content. From the SEM analysis it was proven that when the concentration increased, MWCNTs were uniformly dispersed in the whole matrix. Finally, from the micro-CT it was evident that for MWCNT contents of 3 phr and higher, a homogenous dispersion could be achieved, covering the entire volume of the specimen.
- Conductivity increased with MWCNT content, and at a critical concentration it raised abruptly up to seven orders of magnitude, signifying the transition from the insulating to conductive behavior. Through thermogravimetric analysis, the thermal stability was showed to increase with increasing MWCNT content.
- Variations in conductivity with filler content can be studied in terms of percolation theory, and the determined percolation threshold and critical exponent, at 0.1 Hz and 30 °C, were 1.7 and 2.8, respectively.
- EP3 exhibited improved nanomechanical properties, while increasing concentration further revealed decreasing H and E values. The dispersion of MWCNTs has a great impact on nanomechanical properties, and should be taken into consideration.
- There was a large increase in the stiffness of the system, which resulted in reducing the energy dissipated during the experiment. However, there was more tube–tube slip in the system at higher concentrations, since percolation, coupled to a better dispersion, leads to a higher contact surface among MWCNTs. For the EP5, EP10, and EP15 samples, the dissipated energy increased.
- There was considerable improvement in the wear resistance due to the addition of MWCNTs. Moreover, lower deviation of H/E values revealed the good dispersion in the matrix, with impacts in almost-stable behavior at resistance to wear.

To achieve a better compatibility between MWCNTs and the epoxy matrix, avoiding the formation of agglomerates, and possibly lowering the MWCNT content in the nanocomposite, future work will be focused on the use of MWCNTs with an appropriate surface functionalization. The results can be compared with the main findings of this study, in order to investigate the effect of MWCNT functionalization on the electrical, thermal, and mechanical responses of the nanocomposites for self-sensing applications that can be applied in aerospace. However, this study proved that despite the fact that MWCNTs have not undergone any functionalization or purification process, they can modify the properties of an epoxy resin at a critical concentration, which is a multifunctionality threshold. By this, aerospace structures can be developed using carbon fiber-reinforced composites with MWCNT/nanocomposite epoxy resin, which can lead to enhanced electrical, thermal, and mechanical properties.

Author Contributions: Conceptualization, E.P.K.; methodology, A.-F.A.T., E.P.K., S.G.S., T.G.V.; investigation, A.-F.A.T., E.P.K., S.G.S., T.G.V.; formal analysis, E.P.K., S.G.S., T.G.V.; visualization, A.-F.A.T., E.P.K.; writing—original draft preparation, A.-F.A.T.; writing—review and editing, E.P.K., G.C.P., C.A.C.; supervision, G.C.P. and C.A.C.; project administration, C.A.C.; funding acquisition, C.A.C.

Funding: This research was funded by European Union's Horizon 2020 Research and Innovation Programme MODCOMP (Modified cost effective fibre-based structures with improved multi-functionality and performance), under grant number 685844.

References

1. Tjong, S.C. Structural and mechanical properties of polymer composites. *Mater. Sci. Eng. R Rep.* **2006**, *53*, 73–197. [CrossRef]
2. Thostenson, E.T.; Chou, T.W. Processing–structure multifunctional property relationship in carbon nanotube/epoxy composites. *Carbon* **2006**, *14*, 3022–3029. [CrossRef]
3. Zhang, H.; Zhang, Z.; Zhao, G.; Liu, Y. Influence of carbon black with different concentration on dynamic properties and heat buildup of semi-efficient natural rubber composites. *Micro Nano Lett.* **2016**, *11*, 402–406. [CrossRef]
4. Quilter, A. Composites in Aerospace (White Paper). Available online: https://ihsmarkit.com/pdf/Composites-Aerospace-Applications-whitepaper_264558110913046532.pdf (accessed on 29 December 2018).
5. Ghori, S.W.; Siakeng, R.; Rasheed, M.; Saba, N.; Jawaid, M. The role of advanced polymer materials in aerospace. In *Sustainable Composites for Aerospace Applications*; Woodhead Publishing Series in Composites Science and Engineering; Woodhead Publishing: Sawston, UK, 2018; pp. 19–34.
6. Taraghi, I.; Fereidoon, A.; Mohyeddin, A. The effect of MWCNTs on the mechanical properties of woven Kevlar/epoxy composites. *Steel Compos. Struct.* **2014**, *17*, 825–834. [CrossRef]
7. Selmi, A.; Friebel, C.; Doghri, I.; Hassis, H. Prediction of the elastic properties of single walled carbon nanotubes reinforced polymers. *Compos. Sci. Technol.* **2007**, *67*, 2071–2084. [CrossRef]
8. Giraldo, L.F.; Lopez, B.L.; Bostow, W. Effect of the type of carbon nanotubes on tribological properties of polyamide. *Polym. Eng. Sci.* **2009**, *49*, 896–902. [CrossRef]
9. Gojnya, F.H.; Wichmann, M.H.G.; Köpke, U.; Fiedler, B.; Schulte, K. Carbon nanotube reinforced epoxy composites: Enhanced stiffness and fracture toughness at low nanotube content. *Compos. Sci. Technol.* **2004**, *15*, 2363–2371. [CrossRef]
10. Kao, C.C.; Young, R.J. Assessment of interface damage during the deformation of carbon nanotubes composites. *J. Mater. Sci.* **2010**, *45*, 1425–1431. [CrossRef]
11. Lagoudas, D.C.; Piyush, R.; Thakre, A.; Benzerga, A. *Nanoindentation of CNT Reinforced Epoxy Nanocomposites*; Department of Aerospace Engineering, Texas A & M University: College Station, TX, USA, 2012.
12. Tan, B. Nanomechanics: Aeronautics and Aerospace Prospective. *J. Aeronaut. Aerosp. Eng.* **2012**, *1*, 117. [CrossRef]
13. Sandler, J.K.W.; Kirk, J.E.; Kinloch, L.A.; Shaffer, M.S.P.; Windle, A.H. Ultralow electrical percolation threshold in carbonnanotube epoxy composites. *Polymer* **2003**, *44*, 5893–5899. [CrossRef]
14. ALsafee, A.B.; Al Ajaj, I.A.; Khalil, A. Electrical conductivity of untreated multiwalled carbon nanotube/epoxy composites. *Int. J. Appl. Innov. Eng. Manag.* **2014**, *3*, 22–31.
15. Logakis, E.; Skordos, A.A. Lightning strike performance of carbon nanotube loaded aerospace composites. In Proceedings of the ECCM15—15th European Conference on Composite Materials, Venice, Italy, 24–28 June 2012.
16. Allaoui, A.; El Bounia, N. How carbon nanotubes affect the cure kinetics and glass transition temperature of their epoxy composites?—A review. *Express Polym. Lett.* **2009**, *3*, 588–594. [CrossRef]
17. Meador, M.A. High temperature polymer matrix composites. In *Aeropropulsion '87*; Session 1: Aeropropulsion Materials Research; NASA Lewis Research Center: Cleveland, OH, USA; p. 15.
18. Allaoui, A.; Bai, S.; Cheng, H.; Bai, J. Mechanical and electrical properties of a MWNT/epoxy composite. *Compos. Sci. Technol.* **2002**, *62*, 1993–1998. [CrossRef]
19. Lillehei, P.T.; Kim, J.-W.; Gibbons, L.J.; Park, C. A quantitative assessment of carbon nanotube dispersion in polymer matrices. *Nanotechnology* **2009**, *20*, 325708. [CrossRef]
20. Trompeta, A.F.; Koklioti, M.A.; Perivoliotis, D.K.; Lynch, I.; Charitidis, C.A. Towards a holistic environmental impact assessment of carbon nanotube growth through chemical vapour deposition. *J. Clean. Prod.* **2016**, *129*, 384–394. [CrossRef]
21. Zhou, H.; Han, G.; Xiao, Y.; Chang, Y.; Zhai, H.-J. A comparative study on long and short carbon nanotubes-incorporated polypyrrole/poly(sodium 4-styrenesulfonate) nanocomposites as high-performance supercapacitor electrodes. *Synth. Mater.* **2015**, *209*, 405–411. [CrossRef]
22. May, C.A. Resins for Aerospace. *Appl. Polym. Sci.* **1985**, *285*, 557–580. [CrossRef]
23. Wei, R.; Tu, L.; You, Y.; Zhan, C.; Wang, Y.; Liu, X. Fabrication of crosslinked single-component polyarylene ether nitrile composite with enhanced dielectric properties. *Polymer* **2018**, *161*, 162–169. [CrossRef]

24. High Performance Polymers in the Aerospace and Automotive Sectors, Frost and Sullivan. Available online: https://store.frost.com/high-performance-polymers-in-the-aerospace-and-automotive-sectors.html#section1 (accessed on 29 December 2018).
25. Schilde, C.; Schlomann, M.; Overbeck, A.; Linke, S.; Kwade, A. Thermal, mechanical and electrical properties of highly loaded CNT epoxy composites: A model for the electric conductivity. *Compos. Sci. Technol.* **2015**, *117*, 183–190. [CrossRef]
26. AlQadhi, M.; Merah, N.; Khan, Z.; Mezghani, K.; Gasem, Z.; Adinoyi, M.J. Effect of sonication and high shear mixing parameters on nanoclay dispersion in epoxy. In Proceedings of the ECCM15—15th European Conference on Composite Materials, Venice, Italy, 24–28 June 2012.
27. Liu, C.-X.; Choi, J.W. Improved Dispersion of Carbon Nanotubes in Polymers at High Concentrations. *Nanomaterials* **2012**, *2*, 329–347. [CrossRef] [PubMed]
28. AlQadhi, M.; Merah, N. Method for Preparation of Epoxy Clay Nanocomposites. U.S. Patent 9334387B2, 10 September 2013.
29. Huang, Y.Y.; Ahir, S.V.; Terentjev, E.M. Dispersion rheology of carbon nanotubes in a polymer matrix. *Phys. Rev. B* **2006**, *73*, 125422. [CrossRef]
30. Berkovich, A.R.; Hower, J.C.; Jacques, D.; Rantell, T. Fabrication of Carbon Multiwall Nanotube/Polymer Composites by Shear Mixing. 2001. Available online: https://acs.omnibooksonline.com/data/papers/2001_37.3.pdf (accessed on 30 November 2018).
31. Charitidis, C.A. Nanomechanical and nanotribological properties of carbonbased thin films: A review. *J. Refract. Met. Hard Mater.* **2010**, *28*, 51–70. [CrossRef]
32. Koumoulos, E.P.; Charitidis, C.A.; Papageorgiou, D.P.; Papathanasiou, A.G.; Boudouvis, A.G. Nanomechanical and Nanotribological Properties of Hydrophobic Fluorocarbon Dielectric Coating on Tetraethoxysilane for Electrowetting Applications. *Surf. Coat. Technol.* **2012**, *206*, 3823–3831. [CrossRef]
33. Oliver, W.C.; Pharr, G.M. An improved technique for determining hardness and elastic modulus using load and displacement sensing indentation experiments. *Mater. Res.* **1992**, *7*, 1564–1583. [CrossRef]
34. Sneddon, I.N. Boussinesq's problem for a rigid cone. *Math. Proc. Camb. Philos. Soc.* **1948**, *44*, 492–507. [CrossRef]
35. King, R.B. Elastic analysis of some punch problems for a layered medium. *Int. J. Solids Struct.* **1987**, *23*, 1657–1664. [CrossRef]
36. Bei, H.; George, E.P.; Hay, J.L.; Pharr, G.M. Influence of Indenter Tip Geometry on Elastic Deformation during Nanoindentation. *Phys. Rev. Lett.* **2005**, *95*, 045501. [CrossRef]
37. Kim, H.; Bae, H.; Yu, J.; Kima, S. Thermal conductivity of polymer composites with the geometrical characteristics of graphene nanoplatelets. *Sci. Rep.* **2016**, *6*, 26825. [CrossRef]
38. Bellucci, S.; Balasubramanian, C.; Micciulla, F.; Rinaldi, G. CNT composites for aerospace applications. *J. Exp. Nanosci.* **2007**, *2*, 193–206. [CrossRef]
39. Psarras, G. Charge transport properties in carbon black/polymer composites. *J. Polym. Sci. Part B* **2007**, *45*, 2535–2545. [CrossRef]
40. Psarras, G. Conductivity and dielectric characterization of polymer nanocomposites. In *Polymer Nanocomposites: Physical Properties and Applications*; Woodhead Publishing Limited: Cambridge, UK, 2010; pp. 31–69.
41. Pontikopoulos, P.; Psarras, G. Dynamic percolation and dielectric response in multiwall carbon nanotubes/poly(ethylene oxide) composites. *SAM* **2013**, *5*, 1420. [CrossRef]
42. Saito, S.; Sasabe, H.; Nakajima, T.; Yada, K. Dielectric relaxation and electrical conduction of polymers as a function of pressure and temperature. *J. Polym. Sci. A2* **1968**, *6*, 1297. [CrossRef]
43. Xu, F.; Sun, L.X.; Zhang, J.; Qi, Y.N.; Yang, L.N.; Ru, H.Y.; Wang, C.Y.; Meng, X.; Lan, X.F.; Jiao, Q.Z.; et al. Thermal stability of carbon nanotubes. *J. Therm. Anal. Calorim.* **2010**, *102*, 785–791. [CrossRef]
44. Si Eun, L.; Seho, C.; YoungSeak, K. Mechanical and thermal properties of MWCNT reinforced epoxy nanocomposites by vacuum assisted resin transfer molding. *Carbon Lett.* **2014**, *15*, 32–37.
45. Baldissera, A.F.; de Miranda, K.L.; Bressy, C.; Martin, C.; Margaillan, A.; Ferreira, C.A. Using Conducting Polymers as Active Agents for Marine Antifouling Paints. *Mater. Res.* **2015**, *18*, 1129–1139. [CrossRef]
46. Koumoulos, E.P.; Charitidis, C.A.; Daniolos, N.M.; Pantelis, D.I. Nanomechanical properties of friction stir welded AA6082T6 aluminum alloy. *Mater. Sci. Eng. B* **2011**, *176*, 1585–1589. [CrossRef]
47. Li, Y.; Yu, T.; Pui, T.; Chen, P.; Zheng, L.; Liao, K. Fabrication and characterization of recyclable carbon nanotube/polyvinyl butyral composite fiber. *Compos. Sci. Technol.* **2011**, *71*, 1665–1670. [CrossRef]

48. Zhuang, G.S.; Sui, G.X.; Sun, Z.S.; Yang, R. Pseudoreinforcement effect of multiwalled carbon nanotubes in epoxy matrix composites. *J. Appl. Polym. Sci.* **2006**, *102*, 3664–3672. [CrossRef]
49. Fiedler, B.; Gojny, F.H.; Wichmann, M.H.; Nolte, M.C.; Schulte, K. Fundamental aspects of nanoreinforced composites. *Compos. Sci. Technol.* **2006**, *66*, 3115–3125. [CrossRef]
50. Cheng, Y.T.; Cheng, C.M. What is indentation hardness? *Surf. Coat. Technol.* **2000**, *133–134*, 417–424. [CrossRef]
51. Leyland, A.; Matthews, A. Design criteria for wear resistant nanostructured and glassy metal coatings. *Surf. Coat. Technol.* **2004**, *177–178*, 317–324. [CrossRef]
52. Leyland, A.; Matthews, A. Optimization of Nanostructured Tribological Coatings. In *Nanostructured Coatings*; Springer: New York, NY, USA, 2007.

 *aerospace*

Article

Synergy Effects in Electric and Hybrid Electric Aircraf

Teresa Donateo *,† , **Claudia Lucia De Pascalis** † **and Antonio Ficarella**

Department of Engineering for Innovation, University of Salento, 73100 Lecce, Italy;
claudia.depascalis@unisalento.it (C.L.D.P.); antonio.ficarella@unisalento.it (A.F.)
* Correspondence: teresa.donateo@unisalento.it; Tel.: +39-0832-275-541
† These authors contributed equally to this work.

Received: 3 December 2018; Accepted: 28 February 2019; Published: 6 March 2019

Abstract: The interest in electric and hybrid electric power system has been increasing, in recent times, due to the benefits of this technology, such as high power-to-weight ratio, reliability, compactness, quietness, and, above all, elimination of local pollutant emissions. One of the key factors of these technologies is the possibility to exploit the synergy between powertrain, structure, and mission. This investigation addresses this topic by applying multi-objective optimization to two test cases—a fixed-wing, tail-sitter, Vertical Take-off and Landing Unmanned Aerial Vehicle (VTOL-UAV), and a Medium-Altitude Long-Endurance Unmanned Aerial Vehicle (MALE-UAV). Cruise time and payload weight were selected as goals for the first optimization problem, while fuel consumption and electric endurance were selected for the second one. The optimizations were performed with Non-dominated Sorting Genetic Algorithm-II (NSGA-II) and S-Metric Selection Evolutionary Multiobjective Algorithm (SMS-EMOA), by taking several constraints into account. The VTOL-UAV optimization was performed, at different levels (structure only, power system only, structure and power system together). To better underline the synergic effect of electrification, the potential benefit of structural integration and multi-functionalization was also addressed. The optimization of the MALE-UAV was performed at two different levels (power system only, power system, and mission profile together), to explore the synergic effect of hybridization. Results showed that large improvements could be obtained, either in the first test case when, both, the powertrain design and the aircraft structure were considered, and in the optimization of the hybrid electric UAV, where the optimization of the aircraft flight path gave a strong contribution to the overall performances.

Keywords: electrification; modelling and simulation; multi-objective optimization; design of advanced power systems; VTOL-UAV

---

1. Introduction

The superior energy density of hydrocarbon fuel make them a favorite technology for aircraft propulsion and mobility, in general. However, an ever-increasing interest in the electric systems is encouraged by the remarkable benefits of this technology, such as high power-to-weight ratio, efficiency, reliability, compactness, quietness, and, above all, elimination of local pollutant emissions.

Electrification began to take part in the aerospace industry, first, with the More Electric Aircraft (MEA) concept. Then, electric powertrains were introduced also for propulsive purposes, with hybrid-electric propulsion systems as a mid-term solution to improve the overall fuel economy and reduce the environmental impact. A good review of technologies for hybrid electric aircraft can be found in [1]. The growing interest in this field is attested for by the ever-increasing number of studies concerning the electrification or hybridization of existing power systems, such as that in Reference [2], regarding the hybridization of an existing General Aviation (GA) aircraft. However, technical barriers prevent full-electric propulsion systems to reach values of energy storage density (and, therefore, endurance) comparable with hydrocarbon fuel. For this reason, pure-electric propulsion is still considered unconceivable for large aircraft, such as commercial planes, unless new technologies are developed for batteries. On the contrary, it is more and more frequently used as the aerial vehicle size has been decreasing [3]. However, a study by NASA [4] puts into evidence how the integration between the electric propulsion and vehicle systems has its advantages, also with the existing battery technology. Moreover, new technologies like the multi-functionalization of composite materials [5] might lead to further benefits resulting from the synergy between the structure and the propulsion.

To the authors' knowledge, none of the aircraft design procedures presented in the literature, even for the advanced power systems, take into account the synergic effects of architecture, size of the engine, and the electric components, energy management strategy, mission specification, etc., by using, moreover, existing technologies for batteries. Therefore, the main goal of this investigation was to present a comprehensive methodology for the design of electric and hybrid electric aircraft power systems, based on multi-objective optimization algorithms and backward simulations. The proposed method is explained briefly in Section 2 and has been applied to two test cases (Sections 3 and 4) to show how the performance of complex power systems can be strongly improved by exploiting the synergic effects.

The first test case takes inspiration from a study by Aksugur et al. [6], which proposes a pure-electric, propeller-ducted, fan propulsion system to be used as powertrain on a tail-sitter Vertical Take-off and Landing Unmanned Aerial Vehicle (VTOL-UAV). Aksugur et al. [6] selected a propulsion system, a priori, and then completed the UAV design by optimizing four aircraft design parameters. In this investigation, we considered the same UAV, with the same kind of all-electric hybrid propulsion system, but performed optimization of the propulsive units, together with the aircraft architecture. Moreover, further advantages from the structural integration and multi-functionalization were addressed through a survey of the battery integration technologies proposed in the literature. The synergy between the aircraft power system and the flight profile was not addressed in this first test case, as the VTOL-UAV was used for surveillance missions, without a well-defined and repeatable flight profile.

The second problem considered in this study concerned the optimization of both the hybrid electric power system and the mission profile specifications for a Medium-Altitude Long-Endurance Unmanned Aerial Vehicle (MALE-UAV) [7]. This time, the synergy investigated was between the flight mission profile and the aircraft powertrain. The MALE-UAV considered here had a larger size than the first one and required a longer endurance, which made it unsuitable for a full-electric power system. In the case of hybrid electric power systems, a further degree of freedom was represented by the energy management strategy, i.e., by the specific usage of battery and fuel during the mission, which, in turn, depended on the sizing of batteries, engines, and electric machines. In the present investigation, a rule-based technique was implemented to select the usage of the battery in the different phases of flight, but the rules were optimized, together, with the other design parameters.

As for the optimization algorithm, after an analysis of their performance, the authors selected Non-dominated Sorting Genetic Algorithm-II (NSGA-II) [8] and S-Metric Selection Evolutionary Multiobjective Algorithm (SMS-EMOA) [9] for Test Cases 1 and 2, respectively. Another novelty in the approach proposed here, is the choice of the better algorithm for each test case, according to the nature of the optimization problem, in particular the multi-modality of the Pareto Front. The results presented here are part of a larger study on the performance of different evolutionary algorithms over a series of mathematical and engineering optimization problems but, unlike previous studies, this paper focused on the results of the optimization in the aeronautical field.

2. Methodology

The large number of design parameters and goals needed to exploit synergy in electric and hybrid electric power system, calls for the usage of simplified simulation approaches and appropriate optimization tools.

In the proposed methodology, a backward simulation approach is used. This means that the starting point is the request of propulsive power to fly the aircraft. This information is "worked backward" through the components of the power system and the conversion efficiency of the overall powertrain is obtained by multiplying the efficiency of the components in the series. The power request is appropriately split in case of the components, in parallel (Test Case 2). In this way, it is possible to estimate the overall usage of fuel or electricity.

To implement this method and to include the size of the components in the optimization, scalable and composable sub-models are needed for each component [10]. The efficiency of each element can be calculated, either by the average point method or by a quasi-static approach, as explained by Guzzella et al. [10]. In the first approach, the working points of the power system, along a mission, are lumped into a single representative average operating point. This approach is suitable for a simple power system like the one considered in Test Case 1, while it cannot be used for a hybrid electric power system, as in Test Case 2. In fact, a hybrid electric power system relies on a specific energy management strategy that cannot be taken into account in the average point method. For this reason, the power system of Test Case 2 was simulated with a quasi-static approach, where the aircraft mission was discretized as a time sequence of operating points. For each point, the consumption of energy was calculated, as before, but was dependent on the energy management strategy. The utilization of energy at each time step was then integrated, over the mission, to obtain the deployment of fuel or electricity. The amount of charge in the battery, the available fuel, and the weight of the aircraft were updated at each time step. Using this simulation approach, with a suitable time step, it was possible to take into account a series of time-dependent variable like, for example, the voltage of the battery as a function of its state of charge. Of course, the required computational time was much higher than that in the average operating point and this made the use of a heuristic optimization method, mandatory, for the analysis of synergies.

In this investigation, evolutionary algorithms (EAs) were selected because of their ability to address multi-objective optimizations. Among the EAs analyzed by the authors for constrained optimization problems [11,12], SMS-EMOA [9] was found to the best method in all cases, except when the objective functions showed a strong multimodality. In this case, NSGA-II (Non-dominated Sorting Genetic Algorithm-II) [8] was the best. Therefore, before applying the proposed methodology, the multimodality of the objective functions was checked, for each test case.

The last aspect of the methodology to be clarified was how the synergies were quantified. For each test case, this was performed in three steps:

1. The design variables were grouped according to their role (mission, powertrain, architecture, etc.);
2. Optimization runs were performed by considering (as design parameters) the variables of a group, while using an arbitrary setting for all other parameters;
3. The Pareto fronts of the optimization runs and Point 2 were compared with the results of a "synergy" optimization, where the design parameters of all groups were optimized in a single run.

Of course, the accuracy of the results depends on the correctness of the models. The test cases presented were not aimed at the design of a specific power system but only at presenting the proposed methodology. Therefore, we chose models that could guarantee a compromise between accuracy and a reduced computational time.

3. Test Case 1: The VTOL-UAV

The UAV of the first case was an all-electric aerial vehicle. An electric folding propeller was installed on the nose of the aircraft and used for vertical take-off and landing, hovering, and low-speed transition mode phases. While, an Electric Ducted-Fan (EDF) unit was placed between the aircraft tail surfaces, for providing power, during cruise. Both propulsion units consisted of a battery, an electronic speed control (ESC), and a motor. The operational requirements of the UAV were kept constant and are shown in Table 1.

Table 1. Average operating point for the Propeller and the Electric Ducted-Fan (EDF) units.

Propeller unit	3 min at hover mode at 1 km of altitude
EDF unit	Cruise at $V_{cruise} = 50$ m/s at 1 km of altitude

The optimization problem can be formalized as follows:

$$\left\{ \begin{array}{l} \text{maximize } W_{pay}(\overline{x}) \\ \text{maximize } t_{cruise}(\overline{x}) \end{array} \right. \text{subject to} \left\{ \begin{array}{l} V_{stall} \leq 30 \text{ m/s} \\ V_{cruise} \leq 50 \text{ m/s} \\ V_{cruise} - V_{stall} \geq 3 \text{ m/s} \\ \text{Range} \leq 20 \text{ km} \\ t_{cruise} \geq 30 \text{ min} \\ W_{pay} \geq 0.8 \text{ kg} \\ \text{AR} \geq 4 \\ l_{fuselage} \leq b \end{array} \right. \tag{1}$$

The VTOL-UAV model is presented in Section 3.1. It had a dual-objective optimization, with payload weight (W_{pay}) and cruise time (t_{cruise}) as cost functions. Note that the proposed method belongs to the case of "simulation optimization", i.e., the objective functions were calculated through a simulation code.

The design variable vector $\overline{x}$ included the fourteen variables reported in Table 2, together with their lower and upper bounds. Note that the first four variables defined the group related to the architecture. They were the same as those considered in a study by Aksugur et al. [6], from which the test case was derived. The remaining ten design variables were included to the "powertrain" group, as they characterize the size and operating values of the different components of the power system.

Table 2. Design variable bounds for the optimization.

Design Variable	Group	Lower Bound	Upper Bound
Wing span (b)	Architecture	1 m	2 m
Wing loading (W_0/S)	Architecture	100 N/m^2	220 N/m^2
Horizontal tail arm (l_{HT})	Architecture	0.6 m	1.5 m
Take-off weight (W_0)	Architecture	30 N	100 N
Propeller diameter (D_p)	Powertrain	0.5 m	1.2 m
Propeller blade angle (φ_p)	Powertrain	3 deg	30 deg
EDF fan diameter (D_{EDF})	Powertrain	0.07 m	0.16 m
EDF revolutions per minute (n_{EDF})	Powertrain	20,000 rpm	50,000 rpm
Propeller/EDF motor maximum voltage (U_{mMax})	Powertrain	6 V/6 V	24 V/78 V
Propeller/EDF motor maximum current (I_{mMax})	Powertrain	20 A/5 A	180 A/180 A
Propeller/EDF motor KV value (K_v)	Powertrain	60 rpm/V/1200 rpm/V	360 rpm/V/2100 rpm/V
Propeller and EDF batteries maximum discharge rate (C_{rate})	Powertrain	10 C	95 C
Propeller/EDF battery capacity (C_b)	Powertrain	5 Ah/5 Ah	22 Ah/60 Ah
Propeller/EDF battery number of cells in series (S_{bat})	Powertrain	2 S/2 S	10 S/12 S

Note, also, the large number of constrains that ensured the results to be physically sensible and geometrically feasible, by checking the value of a series of variables, are explained in the section on nomenclature. Other technical limitations to be satisfied in terms of voltage, current, and torque of the electrical devices, and tip speed of the blades were also considered.

The optimization tool used for this problem was NSGA-II (Non-dominated Sorting Genetic Algorithm-II) [8], because of the multi-modality of the objective function W_{pay}. The full flowchart of the optimization was reported in Appendix A.

3.1. Modeling the VTOL-UAV

The dependence of the fitness function on the design variables cannot be expressed as a simple mathematical expression, but is the result of a set of sub-models.

The white and blue arrows of Figure 1 show the energy flows throughout the system, while the orange arrows illustrate the backward approach used for the modeling.

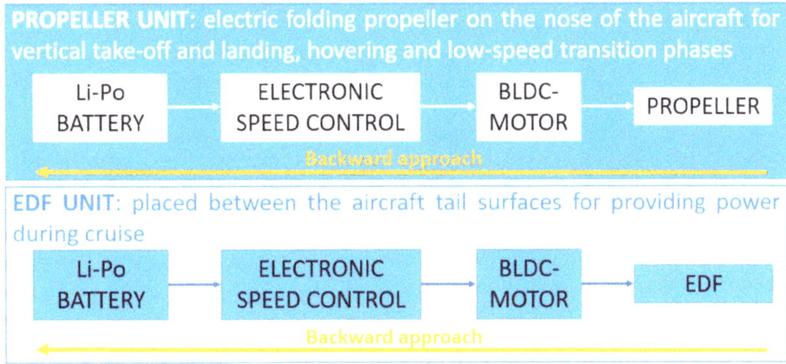

Figure 1. Electric power system workflow.

As explained in Section 2, the required thrust of each propulsion unit is the input of the overall simulation model. The parameters of each component of the power system were adjusted, in order to satisfy the power demand, and the efficiency of each element was taken into account with the average operating point method.

Each sub-model computed the weight of the corresponding element that added to the empty weight of the aircraft (calculated as proposed by Aksugur et al. [6]) and reduced the available payload. In particular, the weight of the propeller and ESC were statistically estimated as a function of their diameter, while the contribution of the motors was statistically estimated as a function of their maximum power. This method took inspiration from Reference [13], where it was possible to find the statistical regression charts of the existing devices. The only component that was not considered in Reference [13] was the EDF, whose weight statistical regression was derived by the authors of this paper and is shown in Figure 2.

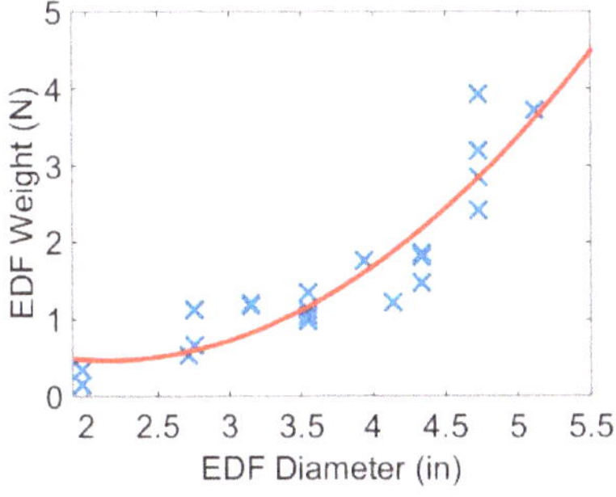

Figure 2. Weight statistical regression for EDF.

3.1.1. Propeller

The propeller was modeled, using Equations (2)–(5). As previously stated, the input parameters of the model were, the required thrust (T_p) and the propeller design parameters. The output required the mechanical power to be calculated as the product of propeller torque and speed.

$$c_T = k_{t0} \cdot N_p \cdot \varphi_p \tag{2}$$

$$c_M = k_{m0} \cdot N_p{}^2 \cdot \left(k_{m1} + k_{m2} \cdot \varphi_p{}^2 \right) \tag{3}$$

$$n_p = 60 \cdot \sqrt{\frac{T_p}{c_T \cdot \rho \cdot D_p{}^4}} \tag{4}$$

$$M_p = c_M \cdot \rho \cdot \left(\frac{N_p}{60} \right)^2 \cdot D_p{}^5 \tag{5}$$

where k_{t0}, k_{m0}, k_{m1}, k_{m2} are parameters depending on the shape and aerodynamics of the propeller blades. In this work, values for the carbon-fiber propellers suggested in Reference [13] were used. Parameters c_T and c_M represent, respectively, the propeller thrust and torque coefficients, while T_p and M_p were the propeller thrust and torque. Finally, n_p represents the revolutions, per minute, and ρ the air density of the flight conditions listed in Table 1.

3.1.2. EDF

The EDF model takes as input parameters, the required thrust and the design variables of Table 2, returning its shaft speed and torque. In particular, the same propeller equations were applied in the EDF model, at first. Then, Equations (6)–(8) were implemented to take into consideration the presence of the duct-wrapping of the fan blades, as suggested in Reference [14].

$$A_{duct} = \pi \cdot \left(\frac{D_{EDF}}{2} \right)^2 \tag{6}$$

$$P_{duct} = P_{unduct} \cdot \sqrt{\frac{A_{unduct}}{2 \cdot A_{duct} \cdot \sigma_{EDF}}} \tag{7}$$

$$M_{shaft} = 60 \frac{P_{duct}}{2 \cdot \pi \cdot n_{EDF}} \tag{8}$$

where, A_{unduct} and A_{duct} are the cross-section areas of the unducted and ducted fans that are assumed to coincide. P_{unduct} and P_{duct} represent the power delivered by the unducted and ducted fan, respectively, σ_{EDF} is assumed to be equal to one and represents the duct diffusion ratio, and M_{shaft} is the torque generated at the EDF shaft.

3.1.3. Motor

The motors considered in this investigation used the Brushless DC technology and were modeled as proposed in [13]. The model takes as input parameters, the design variables described in Table 2, together with the corresponding propulsor (propeller or EDF) revolutions per minute and torque, and calculates the efficiency and the electric power using the following equations:

$$n_{M,max} = \frac{(U_{mMax} - R_M \cdot I_{mMax}) \cdot K_v \cdot U_{M,0}}{U_{M,0} - R_M \cdot I_{M,0}} \tag{9}$$

$$M_{mMax} = \frac{30 \cdot (I_{mMax} - I_{M,0}) \cdot (U_{M,0} - R_M \cdot I_{M,0})}{\pi \cdot K_v \cdot U_{M,0}} \tag{10}$$

$$I_M = \frac{\pi \cdot M_p \cdot K_v \cdot U_{M,0}}{30 \cdot (U_{M,0} - R_M \cdot I_{M,0})} + I_{M,0} \tag{11}$$

$$U_M = R_M \cdot I_{mMax} + \frac{U_{M,0} - R_M \cdot I_{M,0}}{K_v \cdot U_{M,0}} + n_p \tag{12}$$

$$\eta_M = \left(1 - \frac{I_M}{U_M} \cdot R_M\right) \cdot \left(1 - \frac{1}{I_M} \cdot I_{M,0}\right) \tag{13}$$

$$M_M = \frac{\eta_M \cdot I_M \cdot U_M}{2 \cdot \pi \cdot \left(\frac{n_M}{60}\right)} \tag{14}$$

$$P_M = I_M \cdot U_M \cdot \eta_M \tag{15}$$

where $U_{M,0}$ and $I_{M,0}$ represent, respectively, the motor no-load voltage and current, R_M is the motor internal resistance, $n_{M,max}$ and M_{mMax} is the maximum value of revolutions per minute and torque, respectively, while n_p and M_p are the speed and torque of the propulsion unit connected to the electric motor. I_M and U_M are the current and voltage of the motor (defining the electric power requested by the motor), M_M and P_M are the delivered torque and mechanical power, and η_M is the motor efficiency.

3.1.4. ESC

The role of ESC in the considered powertrain is two-fold:

- It has to work as a speed controller for the electric motor after having received the throttle signal from the flight controller.
- According to the working logic of a Brushless Direct Current (BLDC) motor, it transforms the DC current from the battery to a three-phase alternating signal.

The ESC model is described by Equations (16)–(22):

$$U_{ESC,max} = U_{mMax} \tag{16}$$

$$I_{ESC,max} = I_{mMax} \tag{17}$$

$$U_{input} = S_{bat} \cdot 3.7 \tag{18}$$

$$PWF = \frac{I_M U_M}{U_{ESC,max} I_{ESC,max}} \tag{19}$$

$$\eta_{ESC} = (U_{input} - R_{bat} - PTF(1 - PWF))/U_{input} \tag{20}$$

$$P_{ESC} = I_M U_M \tag{21}$$

$$I_{ESC} = \frac{(P_{ESC}/\eta_{ESC})}{U_{input}} \tag{22}$$

where U_{ESC} and I_{ESC} are the ESC nominal voltage and current, $U_{ESC,max}$ and $I_{ESC,max}$ are their maximum values, and P_{ESC} is the power delivered by ESC. U_{input} is the nominal value of the battery voltage that the ESC receives in input, 3.7 is the nominal voltage of a single Lithium-polymer (Li-po) battery cell, and η_{ESC} is the ESC efficiency. Finally, R_{bat} is the battery resistance, PTF represents the part throttle factor, and PWF is the ratio of the actual power to the full power, further details can be found in Reference [15].

Equation (19) is taken from Reference [15] and has been used here to obtain an estimation of the average ESC efficiency that was considered to be constant in time. As pointed out in [15], the ESC efficiency changed with the input voltage U_{input} and the ESC current, during the discharge of the battery. However, this effect was neglected here, because of the simplified nature of the average point method and because the focus of the investigation was not on the actual values of the performance indices. As already explained, the goal of the investigation was to put into evidence some synergic effects in hybrid and electric aircraft.

3.1.5. Battery

The battery model takes as inputs, the ESC efficiency and the electric power of the motor, and returns the battery weight and the discharge-time.

$$P_{bat} = I_M \cdot U_M / \eta_{ESC} \tag{23}$$

$$t_{disch} = \left(\frac{I_{bat}}{C_{bat}} \right)^{1-peuk} \left(\frac{DOD \cdot U_{bat} \cdot C_{bat}}{100 \cdot P_{bat}} \right)^{peuk} \tag{24}$$

$$W_{bat} = \frac{C_{bat} \cdot U_{bat}}{\rho_{bat}} \tag{25}$$

where P_{bat} is the power delivered by the battery, I_{bat} and U_{bat} are the nominal values of the battery current and voltage (which is the product of the nominal cell voltage 3.7 V and the number of cells in series S_{bat}), t_{disch} is the battery discharge time, and W_{bat} is the battery weight. DOD is the depth of discharge of the battery, assumed to equal 80%, while the energy density of the battery, ρ_{bat}, is derived from a database of commercial Lithium-polymer (Li-po) batteries with a capacity upto 22 Ah (see Figure 3, where values between the brackets represent the number of battery cells in series).

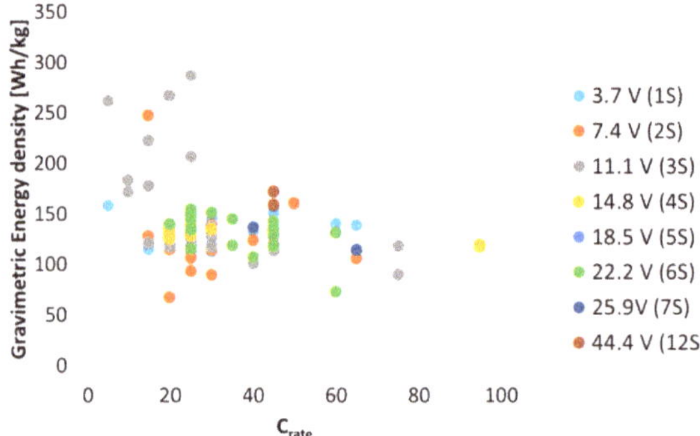

Figure 3. Gravimetric energy density of commercial Lithium-polymer batteries as a function of the battery maximum discharge rate (C_{rate}).

Note that the formula used for the discharge time, i.e., Equation (24), has been extensively validated over a large number of experimental data of constant power discharge of different Li-po batteries, in a previous study by the authors [16]. This formula takes into account the effect of the discharge current on the actual battery capacity, through the Peukert's coefficient "peuk" (set equal to 1.05).

The battery current needs to be checked because it cannot exceed a maximum value $I_{bat,max}$, which depends on its capacity C_{bat} and C_{rate}:

$$I_{bat,max} = C_{bat} \cdot C_{rate} \tag{26}$$

The battery current is checked at the end of the discharge phase, when the cell voltage is assumed to drop to the cut-off value (2.7 V):

$$\frac{P_{bat}}{S_{bat} \cdot 2.7} \leq I_{bat,max} \tag{27}$$

3.2. Results and Discussion

This first test case is used to discuss two different topics. The first one relates to the synergy between the architecture and the power system within the aircraft design proceeding, by using the current technologies for the batteries and motor. The second topic is the further benefit of introducing the multi-functionalization composite materials as a promising technology for improving the overall energy density. Despite both topics having already been addressed in literature, from a qualitative point of view, a quantitative analysis has not yet been performed, to the authors' knowledge.

Synergy between the Architecture and the Power System

The results of the optimization are shown in Figure 4. In particular, Figure 4a compares three Pareto optimal fronts, obtained by different optimizations, using the following as design variables:

- Fixed Arch: Only powertrain specifications (by maintaining a fixed architecture);
- Fixed Pow: Only architecture parameters (fixed powertrain);
- Synergy: Simultaneous optimization of the power system and the architecture of the design variables.

The second optimization allows an indirect comparison with a study by Aksugur et al. [6], as it considers the same inputs but calculates the goals with the models presented here. Note that a direct comparison was not possible because of a lack of information on the values of the design variables and an incomplete description of the numerical procedures.

Note the importance of the introduction of the power system parameters into the aircraft design optimization. In fact, due to the synergy between the aircraft design and powertrain, it was possible to obtain strong enhancements in the target performances (higher payload weight and cruise time). For example, due to the synergy, it was possible to obtain:

- The same payload of 1 kg with a 267% higher cruise time (from about 122 to about 250 min), with respect to the optimization at fixed powertrain.
- The same payload of 1 kg with a 8.2% higher cruise time (from about 414 to about 448 min), with respect to the optimization at a fixed architecture.
- The same cruise time of 130 min with a 228% higher payload weight (from about 0.82 to about 2.69 kg), with respect to the optimization at fixed powertrain.
- The same cruise time of 130 min with a 12.1% higher payload weight (from about 2.40 to about 2.69 kg), with respect to the optimization at fixed architecture.

From Figure 4, it is possible to observe that the main improvements were achieved through the powertrain optimization for a fixed size of aircraft. While, the choice of fixing it, a priori, and optimizing the aircraft architecture, only led to results that provided strongly suboptimal solutions. Finally, synergy allowed to reach the best configurations.

Figure 4b represents a normalized curve of the "Synergy" Pareto front of Figure 4a, and was used to denote three optimal design points, among the optimal solutions of NSGA-II, which were solutions of maximum payload (Max_W_{pay}), maximum cruise time (Max_t_{cruise}), and best trade-off (Best_tradeoff). The latter was chosen by defining the minimum distance of the curve from a so-called "optimum" point having the same cruise time of the solution Max_t_{cruise} and the same payload weight of the design Max_W_{pay}. Therefore, the optimal point had coordinates (1,1) in the normalized plane and (3.51,489.2) in the plot of Figure 4a.

The final choice between the optimal solutions depended on the relative importance of the two goals and was left to the final decision maker. However, their specifications, which are reported in Table 3, could be used to draw some conclusions about the role of the different design parameters on the UAV performances. In particular, it is possible to notice that, as expected, the highest value of the cruise time was reached with an EDF battery capacity that was higher than any other solutions, while reducing the maximum battery discharge rate. This was reflected in the lower EDF motor maximum current and a larger EDF diameter, by maintaining, about constant, the EDF revolutions per minute. This meant that, the motor output torque was reduced, which was probably due to its direct proportionality to the motor maximum current. As for the maximum payload weight, it was possible to notice the strong reduction of the battery capacity. Another parameter that was reduced was the fan diameter, while the EDF motor rpm was about constant, with a consequent increase of the maximum currents, to which the motor torque was directly proportional. Based on these observations, it was possible to conclude that the battery specifications represented the main factor of influence on the final performances. This corroborates the more and more increasing interest in improving their technologies. However, this also study shows that optimization is a precious tool for better exploiting the existing technology's potential. Finally, confirming what has been stated above about the aircraft powertrain and the structure optimization, it was possible to observe that the main differences between the three configurations listed in Table 3 were in some parameters of the power system.

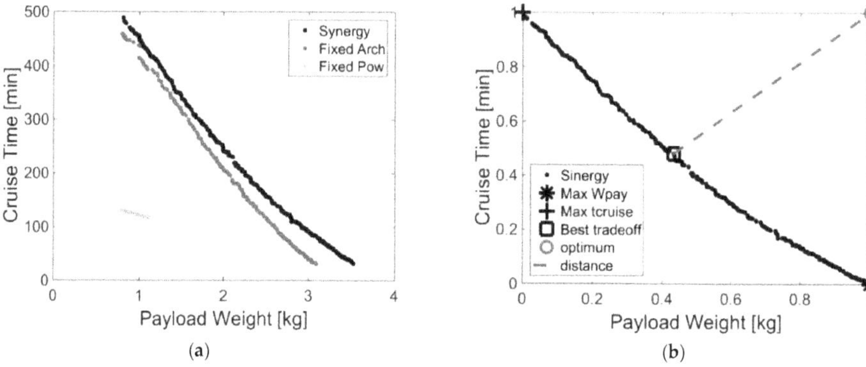

Figure 4. Pareto fronts the VTOL-UAV optimization problem. (**a**) Comparison of the three problems; and (**b**) choice of the optimal solutions.

Table 3. Design variables of the three optimal design points of the VTOL-UAV optimization problem.

Design Point	b (m)	l_{HT} (m)	W_0/S (N/m)	W_0 (N)	D_p (m)	φ_p (deg)
Max_W$_{pay}$	1.78	0.83	219	69	0.77	29.7
Max_t$_{cruise}$	1.73	0.88	220	65	0.76	30.0
Best_tradeoff	1.76	0.94	220	69	0.77	29.9

	U_{mMax} (V) Propeller	I_{mMax} (A) Propeller	K_v (rpm/V) Propeller	U_{mMax} (V) EDF	I_{mMax} (A) EDF	K_v (rpm/V) EDF
Max_W$_{pay}$	18.6	91.3	61.4	15.0	9.3	1804
Max_t$_{cruise}$	18.6	84.1	61.1	14.9	5.7	2052
Best_tradeoff	18.6	89.5	61.2	16.0	6.1	1942

	C_{rate} propeller	C_b (Ah) propeller	C_{rate} EDF	C_b (Ah) EDF	n_f (rpm)	D_f (m)
Max_W$_{pay}$	90C	7.3	82C	6.3	26,090	0.09
Max_t$_{cruise}$	94C	7.0	19C	44.0	25,700	0.15
Best_tradeoff	92C	7.3	73C	27.5	25,530	0.14

	n_p	S_{bat} Propeller	S_{bat} EDF	W_{pay} (kg)	t_{cruise} at V_{max} (min)
Max_W$_{pay}$	4	5	4	3.51	32.6
Max_t$_{cruise}$	4	5	4	0.82	489.2
Best_tradeoff	4	5	4	1.99	252.1

3.3. Structural Integration and Multi-Functionalization

Structural integration means embedding a part into the primary structure of the overall system and making the latter fulfil some structural tasks (such as carrying loads and shape retention), in order to save weight and volume. A step ahead of the structural integration is the multi-functionalization, which consists of designing system components to perform multiple (at least two) predetermined functions. An interesting example is the multi-functionalization of the fiber-reinforced composite materials, to add electrical energy storage capabilities [5]. This allows for part of the conventional energy storage system mass to be subsumed into the aircraft structure, reducing the conventional energy storage system mass needed for a given total amount of stored energy. Therefore, multifunctional composites can be very interesting in the design of electrified airplanes. Note that, the structural properties and failure modes of the modified material should be investigated, as well as the storage ability of the multifunctional component, as a function of shapes, materials, loads, and strains. However, this is beyond the scope of this investigation.

Let us define σ_s and σ_e [5] as:

$$\sigma_e = \frac{\text{actual energy density}}{\text{reference energy density}},$$
$$\sigma_s = \frac{\text{actual specific structural property}}{\text{reference specific structural property}} \tag{28}$$

where superscripts "battery" or "structure" can be added to each term, to specify if the term is related to either the structure or the battery. Note that σ_s represents the structural properties (e.g., stiffness or strength density) and σ_e measures the energy mass efficiency. In the conventional case (functional separation and no-integration), $\sigma_s^{battery} = \sigma_e^{structure} = 0$, and $\sigma_e^{battery} = \sigma_s^{structure} = 1$.

If the structural parts get energy storage tasks and the battery contributes to the structural functions, it is possible to quantify the mass saving potential, as follows [5]:

$$m_{total} = m_{structure} + \frac{m_{battery}}{\sigma_e^{battery}} - \frac{\sigma_s^{battery} m_{battery}}{\sigma_e^{battery}} \tag{29}$$

In Equation (29), m_{total} is the total mass of the system, $m_{structure}$ is the mass of the system structure, and $m_{battery}$ is the conventional battery mass. Note that the first term of Equation (29) represents the original aircraft structure mass, the second term is the mass of the multifunctional battery as a function of its energy mass efficiency, and the mass of the conventional battery. The third term represents the mass to be subtracted to the original structure mass of the aircraft, due to the presence of a multifunction battery that contributes to the structural task.

Let us now consider the further improvements of the VTOL-UAV optimization that could be obtained with the integration/multi-functionalization concept and the technologies described above. This could be performed by identifying the values of $\sigma_s^{battery}$ and $\sigma_e^{battery}$ which:

- starting from the solution Max_t$_{\text{cruise}}$, gave the same payload as that of solution Best_tradeoff, see Figure 5a;
- starting from the solution Best_tradeoff, allowed same payload as that of solution Max_W$_{\text{pay}}$, see Figure 5b;
- starting from the solutions Max_t$_{\text{cruise}}$, reached the same payload as that of solution Max_W$_{\text{pay}}$.

The last case was found to not be feasible. While, the results of the first two cases are shown in Figure 5, where the total mass represents the aircraft's empty weight, in order to get the wanted payload, which is 1.99 kg and 3.51 kg in Figure 5a,b, respectively.

The couples of values that led to obtaining W_{pay} = 1.99 kg, starting from solution Max_t$_{\text{cruise}}$, are shown in Figure 5a and are: ($\sigma_e^{battery}$ = 1,$\sigma_s^{battery}$ = 0.86), ($\sigma_e^{battery}$ = 0.8,$\sigma_s^{battery}$ = 0.88), and ($\sigma_e^{battery}$ = 0.5,$\sigma_s^{battery}$ = 0.93). In this case, the empty weight (or total mass) should be equal to 4.64 kg, while, the couples of values that allowed W_{pay} = 3.51 kg to be reached, starting from the solution "Best_tradeoff", are shown in Figure 5b, and are: ($\sigma_e^{battery}$ = 1,$\sigma_s^{battery}$ = 0.75); ($\sigma_e^{battery}$ = 0.8, $\sigma_s^{battery}$ = 0.80), and ($\sigma_e^{battery}$ = 0.5,$\sigma_s^{battery}$ = 0.87). In this case, the empty weight should be equal to 3.48 kg.

Note that the same result could be obtained with different combinations of the two parameters. In particular, it was possible to observe that $\sigma_e^{battery}$ decreased as the $\sigma_s^{battery}$ increased. Since it is desirable for the battery to maintain most of its energy efficiency when the multi-functionalization concept was applied, giving it structural tasks, a multi-objective optimization aiming at maximizing both the efficiencies could indicate sensible parameters to be focused on, in further studies on the multi-functionalization.

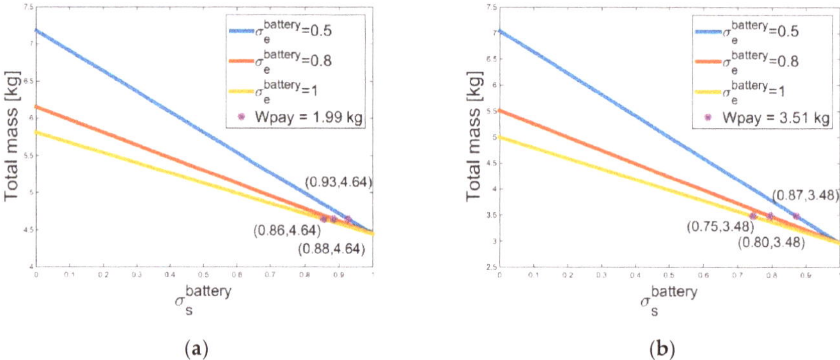

Figure 5. System masses for the various structural efficiencies (σ_s) and Energy Densities (σ_e). (**a**) Payload weight (W_{pay}) = 1.99 kg; and (**b**) W_{pay} = 3.51 kg.

The next step was to verify if the technologies proposed in the literature for the multi-functional panels could reach these requirements. Reference [17] suggests various approaches for creating multi-functional energy storage devices and describes the development of structural batteries, capacitors, and supercapacitors. In addition, Reference [5] provides various studies as examples of different degrees of integration and multi-functionalization. Among the proposed technologies, the storage device shown in Figure 6 seems to be the most suitable for the goals of this investigation. It considers carbon fibers fabric as an anode, a metal substrate as a primary electrical bus (current collector), coated with a thin film of active cathode, glass-fiber mats as separators, and a matrix material for the electrolyte [17,18]. From the results of References [18,18] (Figure 7), it was possible to notice that this technology could provide a high electrical capacity and conductivity, while maintaining good mechanical properties. However, Figure 7 represents the results of the initial analysis, considering standalone materials whose properties were less critical than those of the whole composite. Furthermore, fibers analyses did not consider, either the impact of the electrochemical cycles on the fiber mechanics or the effects of the mechanical load on the electrical properties. Further details can be found in Reference [18].

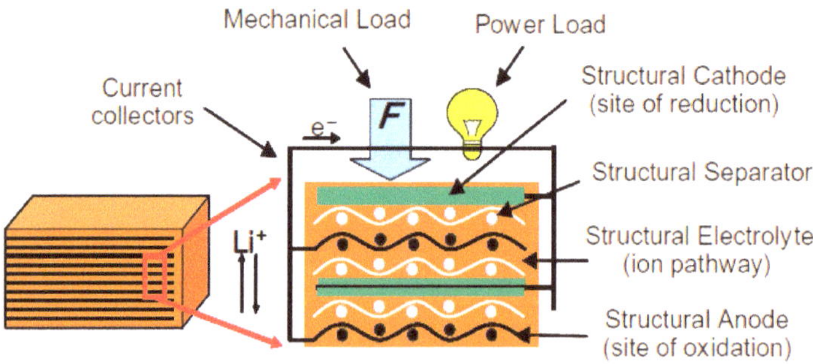

Figure 6. Structural battery scheme.

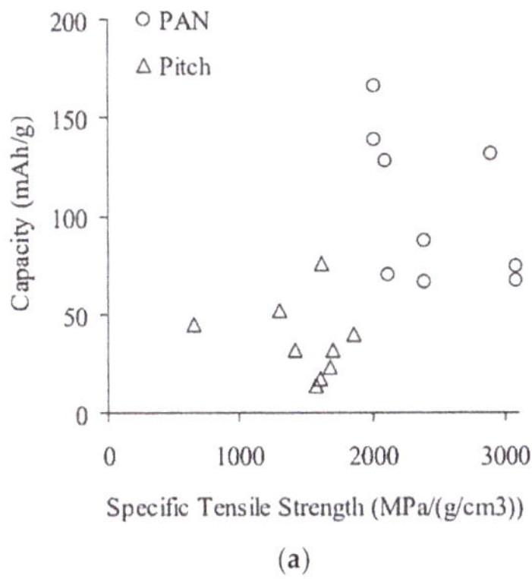

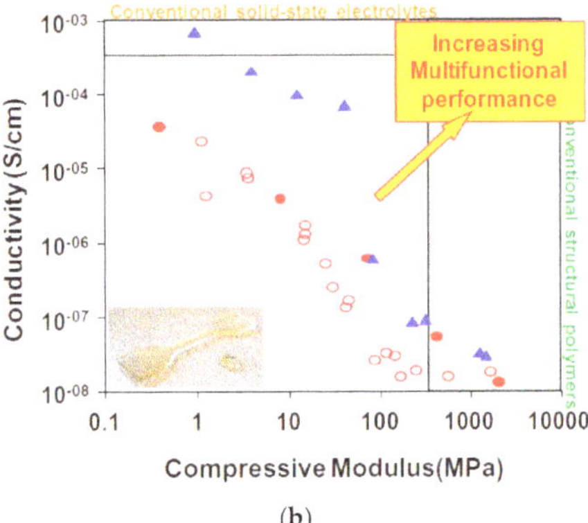

(a)   (b)

Figure 7. Multifunctional technology. (**a**) Multifunctional properties of carbon fibers [17,18]; and (**b**) multifunctional performance of polymer-based structural electrolytes [17]. In (**b**), open circles correspond to homopolymers electrolytes, filled circles are copolymers prepared from the two monomers whose homopolymers yielded the highest conductivity and rigidity, triangles correspond to "gels" comprising a structural polymer and liquid electrolyte in different volume ratios [17].

4. Test Case 2: Hybrid Electric MALE-UAV

The second test case refers to a hypothetical UAV, similar to the well-known General Atomics Predator RQ (wing span 14.8 m, wing area 11.5 m^2, and a take-off mass 1,020 kg). The Predator RQ was equipped with a gasoline engine, while a parallel hybrid electric power system was considered in this study. In particular, a Permanent Magnet electric machine and a two-stroke diesel engine were mechanically connected to the propeller, through an appropriate gear box, as shown in Figure 8. Two energy sources were available—a fuel tank and a battery.

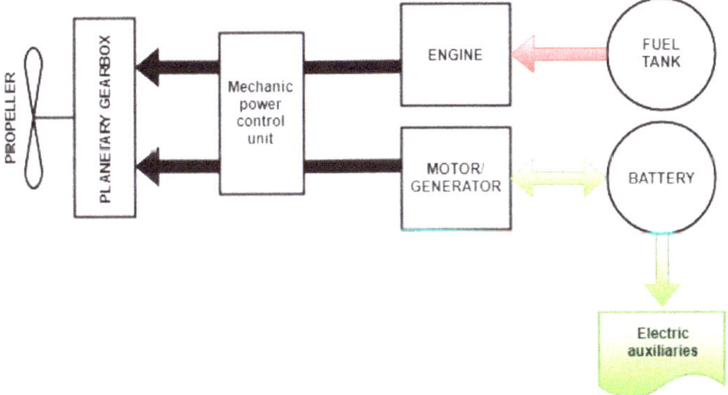

Figure 8. Scheme of the hybrid electric powertrain.

The electric auxiliaries in Figure 8 were assumed to draw a power of 3 kW, at each phase of flight, as suggested in Reference [19].

The objective functions of this problem were the maximum electric endurance of the battery in a single discharge cycle (to be maximized) and the total fuel consumed during the whole mission (to be minimized). These goals were subject to a series of constraints to ensure that:

- The total fuel consumed did not exceed the total fuel mass transported by the airplane, which was a constant value of the problem;

- The battery discharge and recharge currents were lower than their maximum values, C_{rate} and rC_{rate}, respectively, which in turn depended on the battery technology;
- The take-off field length was lower than 1100 m to allow the usage of a large number of runways;
- The additional volume that the hybrid-electric powertrain required was less than 200 L;
- The power system had to guarantee, at each time step, a power $P_{available}$ at least equal to that needed to fly the aircraft ($P_{required}$). This was not obvious because the engine was downsized, with respect to the takeoff requirement and the additional power of the motor was not available when the battery was discharged.

Summing up, it was possible to formalize the optimization problem as follows:

$$\left\{ \begin{array}{l} \text{maximize}\left(\frac{1}{\text{Fuel consumption}(\overline{x})}\right) \\ \text{maximize}(\text{electric endurance}(\overline{x})) \end{array} \right. \quad \text{subject to} \quad \left\{ \begin{array}{l} \text{Fuel consumption} \leq 300 \text{ kg} \\ \left|I_{battery,charging}\right| \leq rC_{rate}\cdot C \\ I_{battery,discharging} \leq C_{rate}\cdot C \\ \text{take} - \text{off field lenght} \leq 1100 \text{ m} \\ \text{Additional volume} \leq 200 \text{ L} \\ P_{required}(t) \leq P_{avalable}(t) \end{array} \right. \tag{30}$$

The MALE-UAV model presented in Section 4.1. The design vector $\overline{x}$ included several parameters that are grouped in the tables that follow, according to their roles. In particular, Table 4 lists the variables defining the size of the main components of the powertrains, namely the engine, the motor, and the battery. For the battery, it was also possible to choose between the different Lithium-based technologies (as explained later), with different ranges of nominal capacity and different energy densities.

Table 4. Design variables defining the size and technology of the powertrain.

Parameters	Unit	Lower Bound	Upped Bound
Nominal Engine power	kW	40	160
Battery elements in series	–	50	100
Battery technology (T)	–	1	3
Battery nominal capacity (C)	Ah	20(T1)	40(T1)
		50(T2)	80(T2)
		100(T3)	150(T3)

The power request could be fulfilled in four possible operating modes:

- Mode 1. Thermal (the engine produced all the power required by the propeller);
- Mode 2. Electric (the propeller shaft power was generated by the motor using the battery as the only energy source);
- Mode 3. Charging (the engine generated the power to move the propeller and to charge the battery, while the electric motor worked as a generator);
- Mode 4. Power-split (both the engine and the motor generated mechanical power that was delivered to the propeller).

The transition from one mode to the other, i.e., the split of power request between the motor and the engine, was controlled by a simple, rule-based strategy. In particular, take-off was always performed in Mode 4. Climb and descent could be performed in Mode 1, 2, 3, or 4 (when feasible). Cruise was performed as a sequence of discharge/recharge process of the battery, i.e., alternating Modes 4 and 3, or Modes 2 and 3. The specific power split during these phases was regulated by the control parameters discharge and the recharge currents of the batteries, respectively (see Table 5).

Table 5. Design variables related to the energy management strategy.

Parameters	Unit	Lower Bound	Upped Bound
Discharge current	C	0.5	20
Recharge current	C	0.1	2
Energy mode during climb	–	1	4
Energy mode during descent	–	1	4

The mission performed by this kind of UAV [19] consists of seven phases—takeoff, climb, outbound to reach the exploration zone, loitering on-station, inbound, descent, and landing to the base station. Figure 9 shows a schematic mission similar to that found in [19]. The mission includes a very long loiter phase of 24 h, which could not be performed with an all-electric power system. This mission was used as baseline case for optimizing the powertrain only.

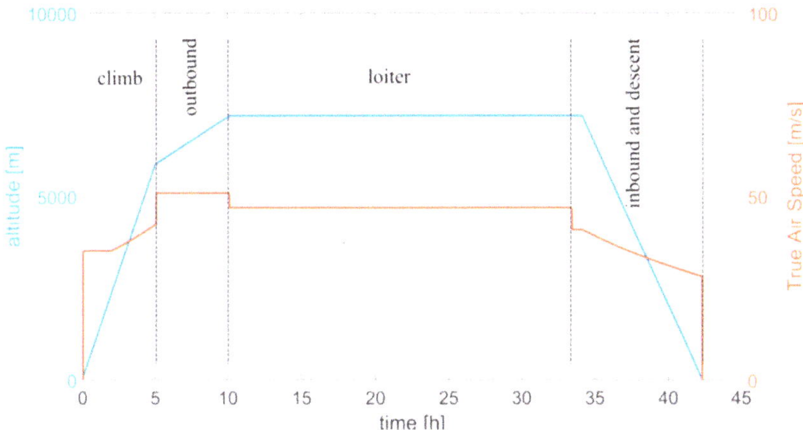

Figure 9. Baseline mission profile for the MALE-UAV.

In the quasi-static simulation approach used in this second test case, it was also possible to include the details of the mission in the optimization, if the designers were given some degrees of freedom in choosing the speeds and altitudes of the different phases of the flight, in particular, the climb, outbound, inbound, and the descent. For this, a parametric mission profile could be generated, using the design variables of Table 6.

Table 6. Mission design variables for the MALE-UAV's optimization.

Parameter	Unit	Min	Max	Step
Outbound speed	m/s	41.0	61.0	0.5
Outbound altitude	m	4900	6900	100
Loiter speed	m/s	37.0	57.0	0.5
Loiter altitude	m	6950	7450	25
Altitude of rate of climb switch	m	2300	3300	25
Rate of climb of the first part of the climb	m/s	0.224	0.424	0.05
Rate of climb of the second part of the climb	m/s	0.224	0.424	0.05
Rate of descent	m/s	0.144	0.344	0.01

Note that this dual-objective optimization has already been discussed in Reference [11], but only in terms of the computational performances of the evolutionary optimization algorithms, to identify the best algorithm for this test case (SMS-EMOA [9]). In the present investigation, the focus was on the results of the optimization and in particular on the synergic effects of mission and powertrain. The full flowchart of this optimization has been reported in Appendix B.

4.1. Simulation Models

Different from the first test case, the performance of the hybrid electric power system, i.e., the two goals, were evaluated as a function of the input variables, with a more detailed simulation approach derived, with appropriate modifications, from the automotive fields [20]. The flowchart of the full simulation process has been shown in Appendix B. Here, follows a brief description of the sub-models used for the components of the powertrain included in the optimization. The propeller was not specifically modeled in this test case, because it was assumed that a propeller with an appropriate efficiency could be selected for each design. The efficiency of the propeller was set to different values, according to the flight phase. Similarly, neither the size of the components nor the energy flows between them were supposed to affect the losses in the mechanical transmission and the coupling device. As in the case of the electric VTOL-UAV, the limited accuracy of these models could affect the absolute values of the performance indices but not the synergic effects analyzed in this paper. For more details on the sub-models and on the quasi-static simulation approach, please refer to [7,20].

4.1.1. Engine

The weight of the engine and its efficiency at each time step were calculated with a scalable model that accounted for engine size, load, and altitude, at a constant speed (Figure 10). The scaling procedure started from a reference engine whose nominal values of power, speed, and brake-specific fuel consumption (bsfc) were known a priori [7], together with the engine weight and the efficiency map (see left part of Figure 10). Then the scaling procedure proposed in [10] was used to estimate the weight, the displacement, and the revolutions per minutes of an engine, with the same technology, but different nominal power (central part of Figure 10). The new engine could be assumed to have the same efficiency map, according to [10]. However, it is well-known that the nominal efficiency of an engine decreases with its size. For this reason an appropriate correction, derived by an interpolation of data of the existing engines, was applied here, to obtain the required information for the scaled engine, as shown in Figure 10 (on the right).

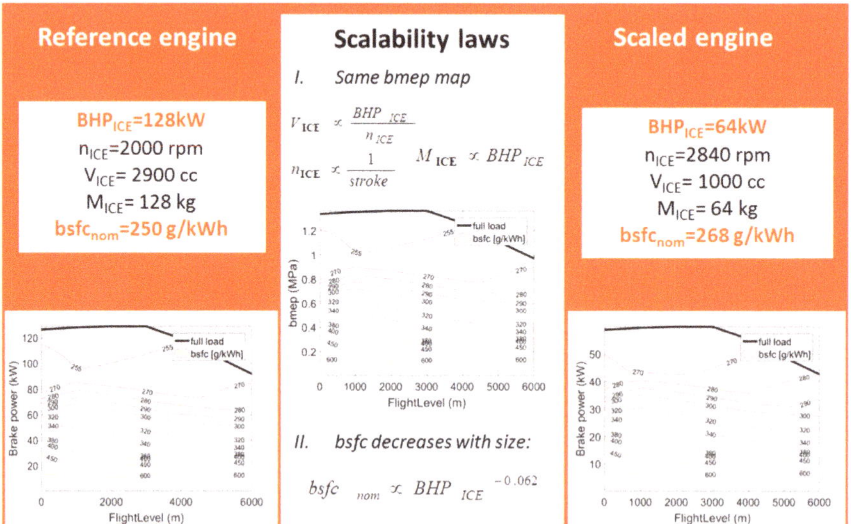

Figure 10. The scalable engine model (BHP_{ICE}: engine brake horse power, n_{ICE}: engine rev/min, V_{ICE}: engine displacement, M_{ICE}: engine mass, bsfc: brake-specific fuel consumption).

4.1.2. Electric Machine

The nominal power of the electric machine was not an input because this component was sized according to the power that the battery could generate in discharge. Its nominal torque (obtained by dividing its power by the fixed speed) was used to predict the nominal efficiency, the mass and the volume of the electric drive (including the motor and the inverter), as better explained in [7]. Again, an interpolation of the data on commercial machines was used.

Along the mission, the electric machine worked as a generator or as a motor, according to the rules of the energy management strategy. Its actual efficiency was calculated as a function of its size and working point with the scalable Willans line method, as suggested in [10]. In the behavior of the generator, the efficiency was mirrored [10]. Therefore, we could write:

$$\eta_{EM} = \left(\frac{P_{mech}}{P_{el}} \right)^{\gamma}$$

(31)

where P_{mech} is the mechanical power, P_{el} the electric power, γ is 1 in motor mode, and -1 in generator mode.

4.1.3. Battery

The mass and the volume of the battery were calculated by interpolating a database of existing research batteries belonging to different technologies (see Acknowledgments), as because commercial database were not available for the large Li-po batteries. These technologies differ from the manufacturing details that were not of interest for this investigation. As a result, the volumetric and gravimetric energy densities of the battery pack were a function of technology and nominal capacity, as in Figure 11. The C_{rate} was 20 C, for all technologies. The recharging could be performed at a maximum current rC_{rate} equal to 2 C.

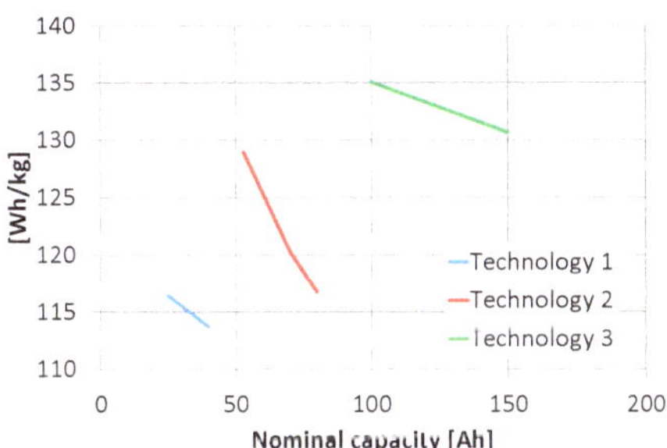

Figure 11. Gravimetric energy density of the researched Li-po batteries with a nominal voltage of 270 V.

At each time step, on the basis of the battery power decided by the energy management strategy (positive in discharge and negative in recharge), the Shepherd's model [21] was used to calculate the open circuit voltage and the current of the battery:

$$V_{OC}(t) = E_0 - \frac{100 \cdot K_b}{SOC(t - \Delta t)} + A \cdot e^{-B \cdot C_{nom} \left(1 - \frac{SOC(t-\Delta t)}{100}\right)}$$

(32)

$$I(t) = \frac{V_{OC}(t) - \sqrt{V_{OC}(t)^2 - 4RR \cdot P_{batt}}}{2RR}$$

(33)

where RR, E_0, A, K_b, and B are parameters of the model that depend on the battery technology and state of charge (SOC) is the percentage of the nominal capacity stored in the battery at the previous time and upgraded after calculating the current with the Coulomb integration method (details in [7]):

$$SOC(t) = SOC(t - \Delta t) - 100 \frac{I(t)}{C_{nom}} \Delta t \tag{34}$$

4.2. Results and Discussion

This problem was used in order to investigate the synergy between the mission profile and the power system, within the powertrain design, with current technologies for engines, batteries, and motors.

4.2.1. Synergy between Mission Profile and Power Systems

The results of the MALE-UAV optimization are shown in Figure 12, where the Pareto optimal fronts were obtained with two different optimizations:

- The overall optimization, called "Synergy", where both the mission profile and the power system parameters are optimized.
- The optimization of the hybrid electric powertrain only, called "Fixed mission", where the mission profile is fixed, a priori.

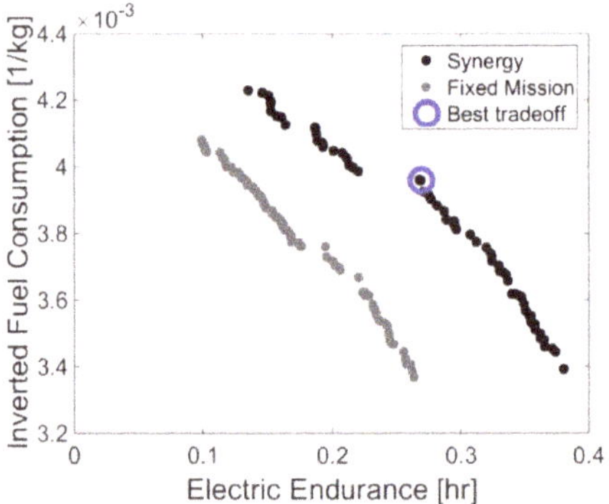

Figure 12. Results of the MALE-UAV optimization problem.

From Figure 12, it is possible to quantify the benefits obtained by exploiting the synergy between the aircraft mission profile and the power system, in terms of both electric endurance and fuel consumption. In particular, if we look at the bounds of the Pareto front obtained, with the two optimizations, we can notice a 44% increase of the maximum electric endurance and a 3.6% reduction of the minimum fuel consumption allowed by the synergy optimization, with respect to the partial "Fixed mission" case. This was because the powertrain components of different sizes operate better in some flight conditions than others, as explained later.

A "Best_tradeoff" solution (also show in Figure 12) was obtained with the same procedure used for Test Case 1. Its mission profile was reported in Figure 13, in terms of both altitude and speed, where it was compared to the baseline mission of Figure 9. While, the values of the other design variables can be found in Table 7.

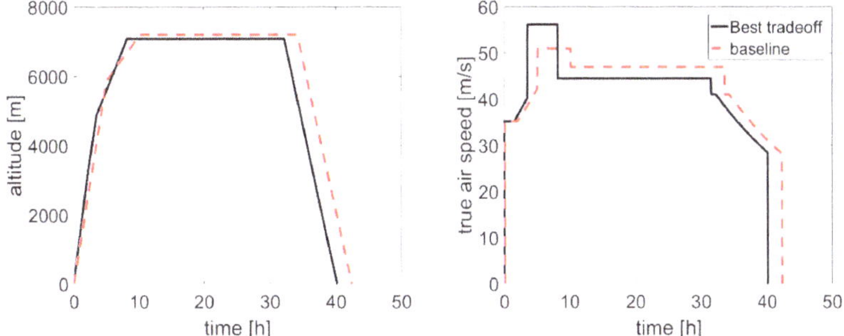

Figure 13. Profile mission of the "Best_tradeoff" solution of Test Case 2.

Table 7. Specification of the "Best_tradeoff" solution of Test Case 2.

Design Point	*DisCur* (C)	*RecCur* (C)	*k*	*Pice,n* (kW)	*Sb*	*Cnom* (Ah)	*BatteryID*
Best_tradeoff	1.03	1.00	2	76	62	53	Li-Po

Compared to the baseline mission of Figure 9, the optimal mission (Figure 13) was characterized by an outbound phase with a higher speed, but a lower altitude, and a faster climb. Note that the loitering-on station was at similar altitude and speed of the baseline case. Since this was the phase of the mission that was specifically connected to the UAV operation, the results of the optimization should not have affected the operational capability of the UAV, but only the time to reach the mission zone (outbound), and come back to the base.

The higher power request of the outbound phase in the optimized mission had a positive effect on the overall fuel economy, because the engine worked at a higher load and, therefore, had a better efficiency. Note that this effect could only be taken into account with the usage of a detailed engine map allowed by the quasi-static approach. Moreover, the optimal mission was shorter than the baseline and this also had a role in the better performance of the tradeoff design. These considerations were confirmed by running the best tradeoff design, over the baseline mission (Figure 14).

This figure also explained the proposed rule-based energy management, consisting of the performance of the loiter phase, as a sequence of discharge (Mode 2) and recharge (Mode 3) phases for the battery, with the engine working at an almost constant load to improve its efficiency. Note that the engine was assumed to not consume fuel, when the powertrain was in Mode 2. However, the fuel consumption associated with the engine restart or idling could easily be included in the modeling approach [10]. Other more complex strategies could be considered, but this was beyond the goal of the present investigation.

It is worthy of note that the mission parameters did not change much over the final "synergy" Pareto front. Therefore, we could conclude that once the mission had been optimized, the trade-off between the overall fuel consumption and electric endurance depended on the choice of the other type of design parameters, namely the size of the components (Table 4), and the energy management (Table 5).

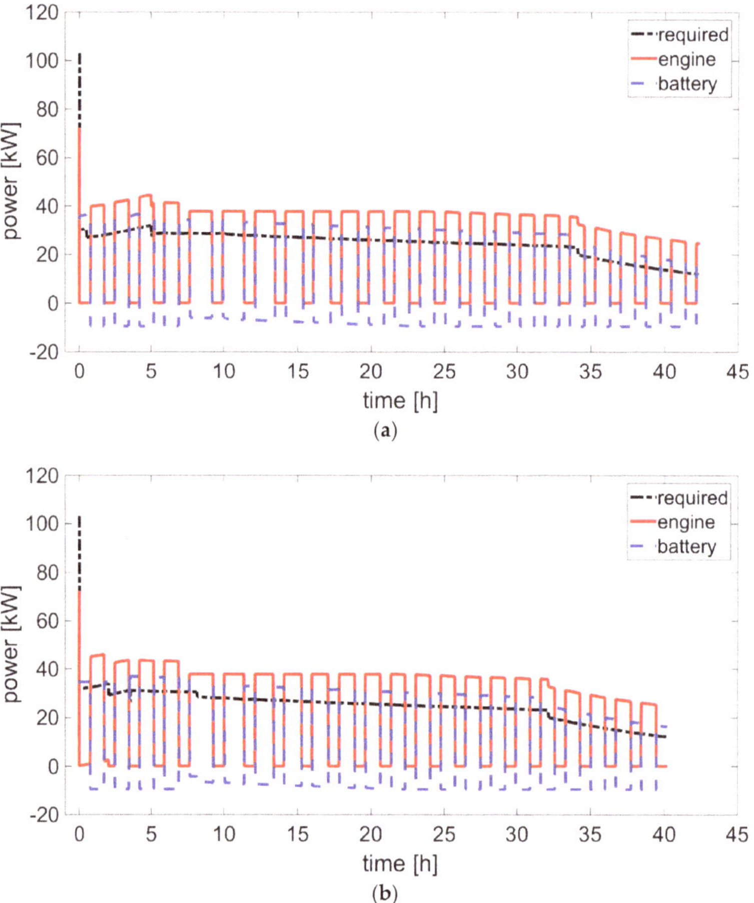

Figure 14. Usage of engine and battery in the optimal tradeoff solution. (**a**) Optimal mission; and (**b**) baseline mission.

4.2.2. Synergy between the Size of the Components and Energy Management

Let us analyze, in details, the variation of the main design parameters, along the fronts in the arbitrary unit, so that the different parameters can be plotted in the same chart. Figure 15 shows how the battery energy and power changes along the fronts of the two optimizations, from the design characterized by a good fuel economy (but poor electric endurance), towards the maximization of electric endurance (at the expense of fuel economy).

The plots shown in the upper part of Figure 15 include the battery nominal energy, which is the product of its nominal capacity and its voltage; also shown in the first row. We can notice that electric endurance was directly correlated to the energy stored in the battery, as expected. However, from left to right, the increase of battery energy was obtained, first, at a constant voltage, by an increasing capacity until a change in battery technology was needed (gap). In fact, the trade-off fronts in Figure 12 show an apparent discontinuity (in correspondence of inverted fuel consumption of about 3.75×10^{-3} kg^{-1}, at a fixed mission, and 4×10^{-3} kg^{-1} in the full optimization), caused by the transition from battery technology 1 to battery technology 2 (Figure 11). After the gap, a further increasing of endurance was obtained at a constant capacity, by increasing the voltage. This was particularly evident in the synergy optimization.

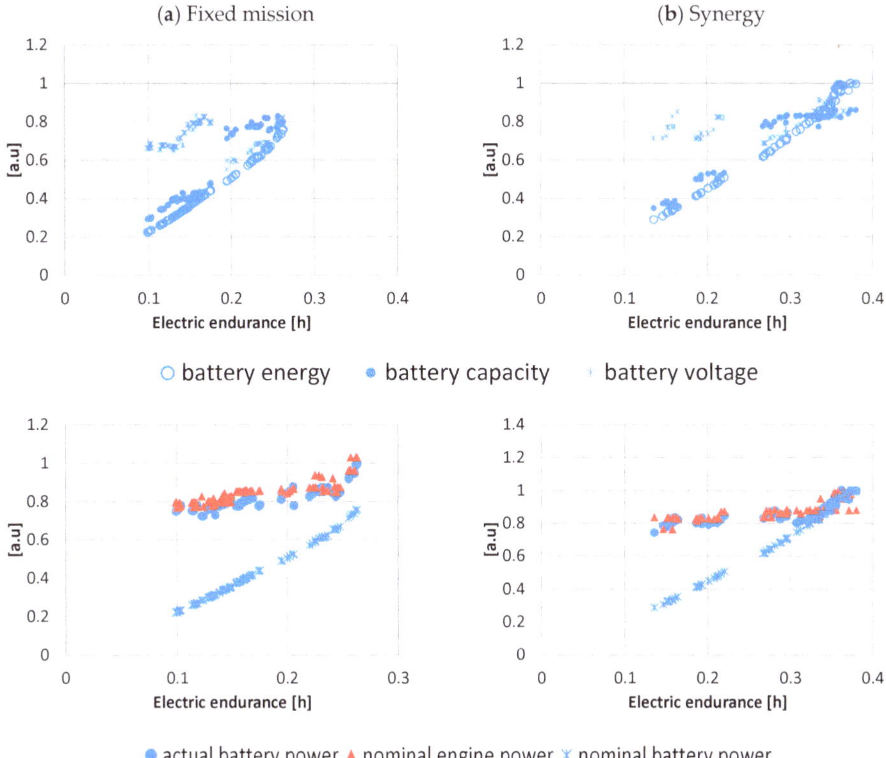

Figure 15. Trends of engine size and battery specifications along the Pareto front, for the fixed mission optimization (on the left) and synergy optimization (on the right).

The plots in the lower part of Figure 15 show the trend of engine and battery powers. For the battery, two kinds of powers are displayed—the nominal power, which is directly proportional to the battery energy, and the actual battery power that depends on the specific energy management strategy. Along the front, from left to right, we can notice a flat course of both engine power and actual battery power, while its energy increased, i.e., to increase the electric endurance, the energy stored in the battery was made increasingly larger but it was discharged at a lower rate. Even if the engine power was the about the same, along a large part of the front, the fuel economy improved because the engine worked at a higher load, due to the increase of the powertrain mass with battery energy. However, this was only possible until a certain value of endurance was reached (0.25 h and 0.34 h in the fixed mission and synergy optimization, respectively). A further improvement of the electric endurance could only be obtained with an increase in both engine power and actual battery power.

5. Conclusions

This investigation proposed a methodology to quantify the synergies between power system, architecture, mission, energy management, and multi-functionalization in electric and hybrid electric unmanned aerial vehicles, with advanced multi-objective optimization tools and appropriate simplified models. To explain the potential of the methodology to exploit the best existing technologies for such advanced power systems (in particular batteries and motors), two test cases were analyzed and discussed.

In the case of the fixed-wing tail-sitter VTOL-UAV, the results show that it was possible to obtain large improvements, in terms of the target performances, due to the synergy between aircraft design and powertrain, with respect to designing structure and power system, separately. In particular, the powertrain optimization gave the main contribution to the results, while, the choice of fixing it, a priori, and optimizing the aircraft architecture, only led to strong suboptimal solutions. In fact, the simultaneous optimization of both architecture and power system provided a 218% higher payload weight, and a 274% higher cruise time, with respect to the optimization of the only aircraft architecture, a +14% payload, and a +7% cruise time, when compared to the optimization of the only powertrain. The levels of multi-functionalization needed to obtain the maximum endurance with the same weight of the trade-off solution, and the maximum payload with the tradeoff endurance were also calculated, but constrained, again, to the existing technologies for multi-functionalization, as found in literature.

Another problem addressed in this work was the simultaneous optimization of the parallel-hybrid electric power system and the mission profile of a Medium Altitude-Long Endurance Unmanned Aerial Vehicle (MALE-UAV), in order to minimize the overall fuel consumption and maximize the electric endurance. An important outcome of this problem is the quantification of the benefits that can be obtained from the exploitation of the synergy between mission profile, the power system, and energy management, in designing a hybrid aircraft by exploiting the engine and the battery, as much as possible.

Since all results were obtained with the existing technologies, the improvements in electric endurance and fuel economy found in this study could possibly be implemented now and not only after future technology developments. Note also that the actual values of the objective functions were affected by the accuracy of the models used for the simulation. However, this did not compromise the quantification of the synergic effects.

Author Contributions: T.D.: Conceptualization, Methodology, Resources, Supervision, Writing, Validation; C.L.D.P.: Formal analysis, Investigation, Resources, Software, Writing; A.F.: Supervision, Funding, Project administration.

Funding: This research was funded by the Italian Ministry for Education, University and Research.

Acknowledgments: The database of research batteries and electric machines used in the second test case were supplied by the partners of a research project of the Aerospace Technological District (DTA-Scarl), funded by the Italian Ministry for Education, University and Research (project code PON03_00067_8).

Conflicts of Interest: The authors declare no conflict of interest.

Nomenclature

c_M	Propeller torque coefficient
c_T	Propeller thrust coefficient
k_{t0}, k_{m0}, k_{m1}, k_{m2}	constant parameters depending on shape and aerodynamics of the propeller blades
σ_e	Energy efficiency
σ_s	Structural efficiency
A	Cross-section area
AR	Aspect ratio
b	Wing span
BatteryID	Typology of battery
BLDC-Motor	Brushless DC Electric Motor
C	Battery capacity
Cnom	Nominal capacity of battery
C_{rate}	Battery maximum discharge rate
D	Diameter
D_f	Diameter of the EDF
DisCur	Current of discharge
DOD	Depth of Discharge
EDF	Electric Ducted Fan
ESC	Electronic Speed Control
I	Current

I_{other}	Current for auxiliaries
k	Climb energy mode
K_v	Motor rpm per volt
$l_{fuselage}$	Fuselage length
l_{HT}	Horizontal tail arm
MEA	More Electric Aircraft
$m_{structure}$	UAV structure mass
m_{total}	UAV total mass
n_{EDF}	Revolutions per minute of the EDF
N_p	Number of propeller blades
P	Power
peuk	Peukert coefficient
$P_{ice,n}$	Nominal power of thermal engine
R	Electrical resistance
Range	Cruise range
RecCur	Current of recharge
ROC	Rate of Climb
ROD	Rate of Descent
S	Wing area
S_{bat}	Number of battery cells in series
T	Thrust
t_{cruise}	Cruise time
t_{dish}	Battery discharge time
U	Voltage
UAV	Unmanned Aerial Vehicle
V_{cruise}	Cruise speed
Vloiter	Loiter speed
Vstall	Stall speed at cruise
VTOL	Vertical Take-Off and Landing
W_0	Take-off weight
W_{pay}	Payload weight
Z	Altitude
η	Efficiency
ρ	Air density
φ_p	Propeller blade angle
$(\cdot)_p$	Propeller property
$(\cdot)_{bat}$	Battery property
$(\cdot)_{EDF}$	EDF property
$(\cdot)_{ESC}$	ESC property
$(\cdot)_{shaft}$	EDF shaft property
$(\cdot)_{duct}$	Ducted fan property
$(\cdot)_{unduct}$	Unducted fan property
$(\cdot)_M$	Electric motor property
$(\cdot)_{max}$	Maximum value
$(\cdot)_{min}$	Minimum value
$(\cdot)_0$	No-load property

Appendix A. Optimization Flowchart of the Electric VTOL-UAV

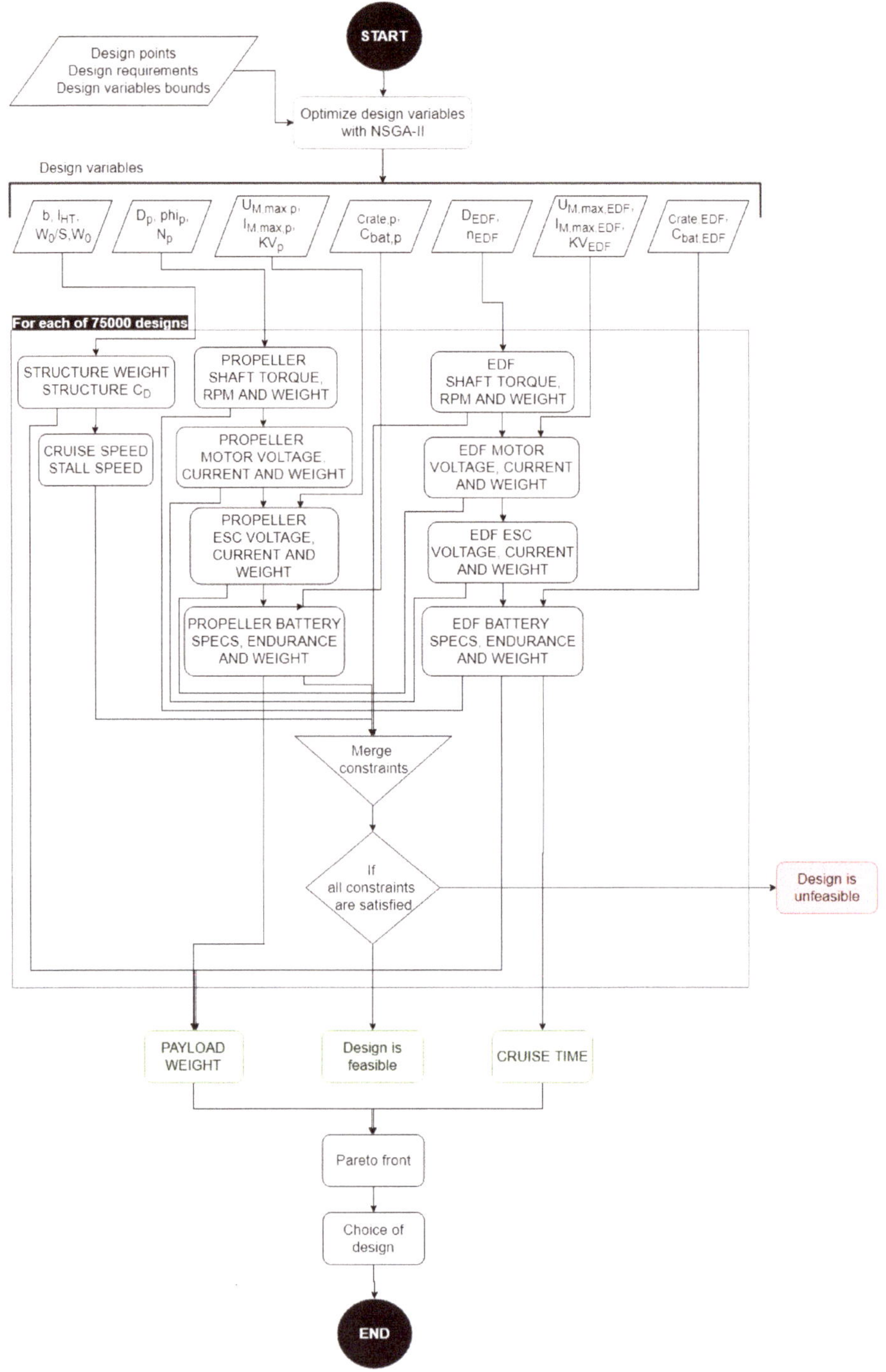

Appendix B. Optimization Flowchart of the Hybrid Electric MALE-UAV

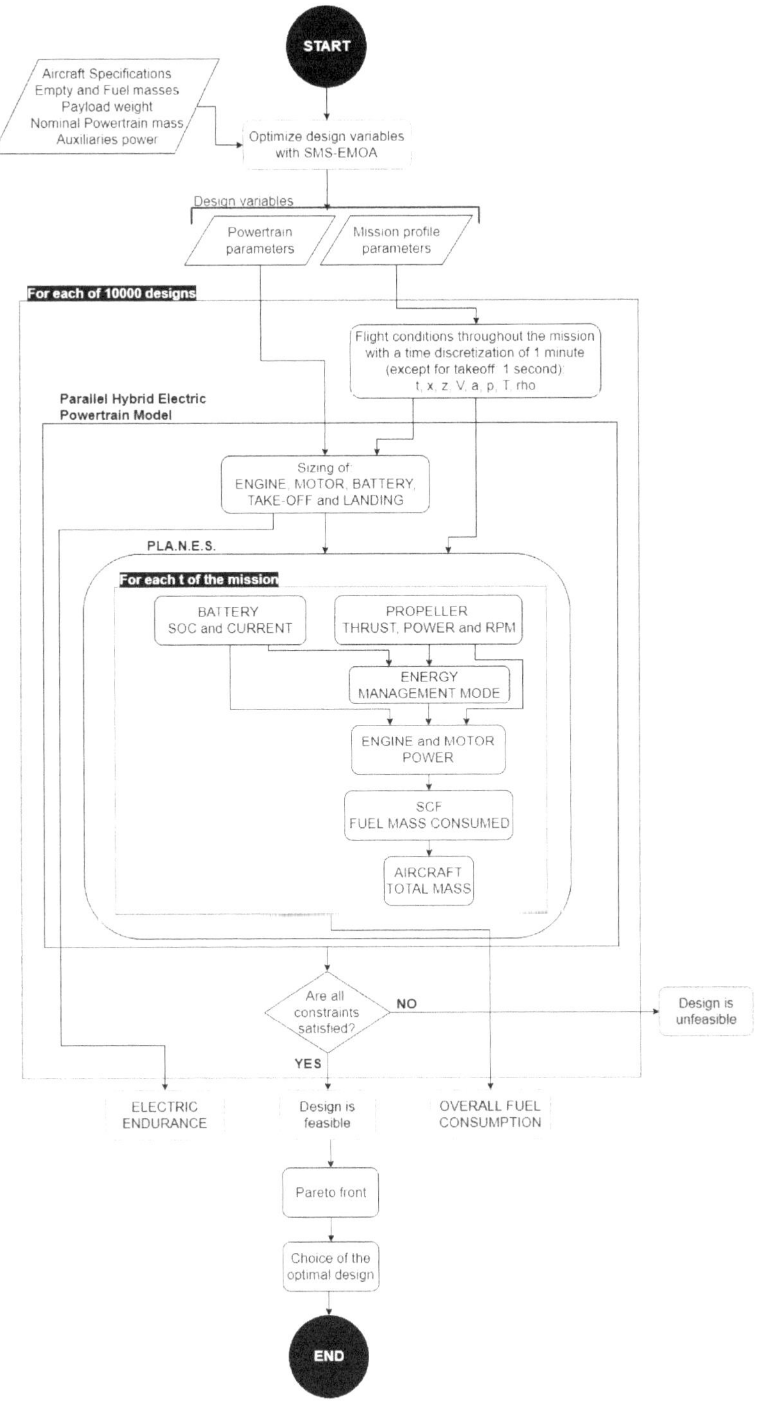

References

1. Abdul Sathar Eqbal, M.; Fernando, N.; Marino, M.; Wild, G. Hybrid Propulsion Systems for Remotely Piloted Aircraft Systems. *Aerospace* **2018**, *5*, 34. [CrossRef]
2. Frosina, E.; Senatore, A.; Palumbo, L.; Di Lorenzo, G.; Pascarella, C. Development of a Lumped Parameter Model for an Aeronautic Hybrid Electric Propulsion System. *Aerospace* **2018**, *5*, 105. [CrossRef]
3. Traub, L.W. Range and endurance estimates for battery-powered aircraft. *J. Aircr.* **2011**, *48*, 703–707. [CrossRef]
4. Moore, M.D. Misconceptions of Electric Aircraft and their Emerging Aviation Markets. In Proceedings of the 52nd Aerospace Sciences Meeting, AIAA SciTech Forum, National Harbor, MD, USA, 13–17 January 2014. AIAA 2014-0535.
5. Adam, T.J.; Liao, G.; Petersen, J.; Geier, S.; Finke, B.; Wierach, P.; Kwade, A.; Wiedemann, M. Multifunctional Composites for Future Energy Storage in Aerospace Structures. *Energies* **2018**, *11*, 335. [CrossRef]
6. Aksugur, M.; Inalhan, G. Design, build and flight testing of a VTOL tailsitter unmanned aerial vehicle with hybrid propulsion system. In Proceedings of the Ankara International Aerospace Conference, Ankara, Turkey, 14–16 September 2011.
7. Donateo, T.; Ficarella, A. Designing of a Hybrid Electric Powertrain for an Unmanned Aircraft with a Commercial Optimization Software. *SAE Int. J. Aerosp.* **2017**, *10*. [CrossRef]
8. Deb, K.; Pratap, A.; Agarwal, S.; Meyarivan, T. A fast and elitist multiobjective genetic algorithm: NSGA-II. *IEEE Trans. Evolut. Comput.* **2002**, *6*, 182–197. [CrossRef]
9. Emmerich, M.; Beume, N.; Naujoks, B. An EMO Algorithm Using the Hypervolume Measure as Selection Criterion. In Proceedings of the Third International Conference of Evolutionary Multi-Criterion Optimization EMO 2005, Guanajuato, Mexico, 9–11 March 2005.
10. Guzzella, L.; Sciarretta, A. *Vehicle Propulsion Systems: Introduction to Modeling and Optimization*; Springer: Berlin, Germany, 2007.
11. Donateo, T.; De Pascalis, C.L.; Ficarella, A. Many-Objective Optimization of Mission and Hybrid Electric Power System of an Unmanned Aircraft. *Lect. Notes. Comput. Sci.* **2018**, *10784*, 231–246. [CrossRef]
12. Donateo, T.; De Pascalis, C.L.; Ficarella, A. Exploiting the synergy between aircraft architecture and electric power system in unmanned aerial vehicle through many-objective optimization. *Int. J. Sustain. Aviat.* **2018**, in printing.
13. Shi, D.; Dai, X.; Zhang, X.; Quan, Q. A Practical Performance Evaluation Method for Electric Multicopters. *IEEE ASME Trans. Mechatron.* **2017**, *22*, 1337–1348. [CrossRef]
14. Jiang, Y.; Zhang, B.; Huang, T. CFD Study of an Annular-Ducted Fan Lift System for VTOL Aircraft. *Aerospace* **2015**, *2*, 555–580. [CrossRef]
15. Gong, A.; Verstraete, D. Development of a dynamic propulsion model for electric UAVs. In Proceedings of the 7th Asia-Pacific International Symposium on Aerospace Technology, Cairns, Australia, 25–27 November 2015.
16. Donateo, T.; Spedicato, L. Fuel economy of hybrid electric flight. *Appl. Energy* **2017**, *206*, 723–738. [CrossRef]
17. Snyder, J.F.; O'Brien, D.J.; Wetzel, E.D. Structural Batteries, Capacitors and Supercapacitors. *Mater. Energy Handb. Solid State Batter.* **2015**, 657–699. [CrossRef]
18. Snyder, J.F.; Baechle, D.M.; Wetzel, E.D.; Xu, K. *Multifunctional Structural Composite Batteries for U.S. Army Applications*; U.S. Army Research Laboratory: Adelphi, MD, USA, 2008.
19. Gundlach, J.F. Multi-Disciplinary Design Optimization of Subsonic Fixed-Wing Unmanned Aerial Vehicles Projected Through 2025. Ph.D. Dissertation, Virginia Polytechnic Institute and State University, Blacksburg, VA, USA, 2004.
20. Donateo, T.; Ficarella, A.; Spedicato, L. A method to analyze and optimize hybrid electric architectures applied to unmanned aerial vehicles. *Aircr. Eng. Aerosp. Technol.* **2018**, *90*, 828–842. [CrossRef]
21. Tremblay, O.; Dessaint, L.-A.; Dekkiche, A.I. A Generic Battery Model for the Dynamic Simulation of Hybrid Electric Vehicles. In Proceedings of the IEEE Vehicle Power and Propulsion Conference, Airlington, TX, USA, 9–12 September 2007.

Article

Synthesis, Analysis, and Design of a Novel Mechanism for the Trailing Edge of a Morphing Wing

Harun Levent Şahin *  **and Yavuz Yaman**

Department of Aerospace Engineering, METU, Ankara 06800, Turkey; yyaman@metu.edu.tr
* Correspondence: harun.sahin@metu.edu.tr; Tel.: +90-555-888-5005

Received: 9 November 2018; Accepted: 7 December 2018; Published: 11 December 2018

Abstract: In the design and analysis of morphing wings, several sciences need to be integrated. This article tries to answer the question, "What is the most appropriate actuation mechanism to morph the wing profile?" by introducing the synthesis, analysis, and design of a novel scissor-structural mechanism (SSM) for the trailing edge of a morphing wing. The SSM, which is deployable, is created via a combination of various scissor-like elements (SLEs). In order to provide mobility requirements, a four-bar linkage (FBL) is assembled with the proposed SSM. The SSM is designed with a novel kinematic synthesis concept, so it follows the airfoil camber with minimum design error. In this concept, assuming a fully-compliant wing skin, various types of SLEs are assembled together, and emergent SSM provide the desired airfoil geometries. In order to provide the required aerodynamic efficiency of newly-created airfoil geometries and obtain pressure distribution over the airfoil, two-dimensional (2D) aerodynamic analyses have been conducted. The analyses show similar aerodynamic behavior with the desired NACA airfoils. By assigning the approximate link masses and mass centers, the dynamic force analysis of the mechanism has also been performed, and the required torque to drive the newly-developed SSM is estimated as feasible.

Keywords: morphing wings; kinematic synthesis; kinematic analysis; scissor-structural mechanisms; scissor-like elements; aerodynamic analysis; dynamic force analysis

---

1. Introduction

Since the early ages, nature has been the main source of inspiration and imagination for humankind due to its grace, complexity, beauty, and mystery in order to solve complex engineering problems. For instance, the flight action varies according to different atmospheric conditions and desired flight paths as hovering, gliding, soaring, and flapping, which are not accomplished by aircraft, but actually by birds by changing their wings rapidly into various forms [1,2].

Researchers had a common sense that aircraft can achieve greater efficiency and productivity if they act like birds. In other words, the analogy with the dynamics of a bird wing requires that the morphing wings eliminate the conventional control surface effects in order to ensure that the flow remains smooth and to minimize the disruption of the surface dislocations and reduce the formation of vortices caused by lift-induced drag [3]. Because of the latest advances in materials science, actuation mechanisms, and structural and manufacturing technologies, the "morphing technology" allows aircraft to use a wide range of wing configurations in flight. First of all, a morphing wing will produce optimum aerodynamic performance over the operational envelope of an aircraft and expand its operating envelope [4,5]. Moreover, by replacing the conventional surfaces with morphing surfaces, the fight control and maneuverability can be improved [6,7]. In addition to the efficient cruise and aggressive maneuvers, the flight range will increase, which will reduce the operational costs by significant fuel savings due to the reduced drag and enhanced thrust generated [8]. Lastly, the use of morphing wings will be expected to play an important role in vibration reduction and will give the opportunity to control flutter, which will significantly improve the comfort and safety and reduce fatigue.

Studies related to the morphing wing can be classified in terms of dimensions that undergo substantial changes such as: planform alternation, airfoil adjustment, and out-of-plane transformation [9]. In the planform alternation, the aircraft wing is aimed at being altered in terms of the span change, the chord-length change, and the sweep-angle change. In the airfoil adjustment category, resizing the thickness and changing the camber-rate of the airfoil is the main purpose. In the out-of-plane transformation, the span-wise and chord-wise bending with wing twisting are intended to be applied [2,9]. It is widely known that the camber of an airfoil has a significant impact on the aerodynamic forces generated under fluid flow [10,11]. Therefore, it is believed that the most effective way to control the forces and moments that occur on aircraft wings is to change the camber-rate of the airfoil [12].

Although the study of the morphing wings requires the simultaneous application of several sciences, researchers have work in specific research areas in order to develop new technologies [13]. Hence, in this article, only the kinematic synthesis of an actuation mechanism, which morphs the trailing edge of an aircraft wing, is extensively investigated. It is assumed that the wing has a compliant skin; therefore, it can undergo any desired displacement.

Starting from the first controlled flight by the Wright brothers, several researchers have proposed the camber/decamber alteration systems for the aircraft wing and control surfaces [14–19]. More recently in 1999, Monner et al. [20] proposed a flexible flap system for an adaptive wing, which varies both in a chordwise and a spanwise differential camber during flight. In 2004, Bartley-Cho et al. [21] addressed the development of smart technologies and the demonstration of the high-rate actuation of hingeless, spanwise, and chordwise deformable control surfaces using a smart material-based actuator in order to improve the aerodynamic performance of a military aircraft. In 2005, Campanile and Anders [22] presented the "belt-rib concept" for variable-camber airfoils, which was developed at DLR (German Aerospace Centre) in the framework of the Adaptive Wing project (in German: Adaptiver Flügel—ADIF). In that concept, instead of using articulated mechanisms, the belt-rib concept is implanted, which can be actuated by piezoceramics or shape memory alloys. In 2009, Marques et al. [23] suggested a variable camber flap concept, which resulted in significant drag reduction and energy savings compared to conventional flaps. In 2011, Vos and Barrett [24] used a pressure-adaptive honeycomb in the design of the trailing edge of a morphing wing. In 2016, Takahashi et al. [25] developed a variable-camber morphing wing composed of corrugated structures. In 2016, Pecora et al. [26] presented a novel wing flap, which enables bi-modal airfoil camber morphing. In 2017, Wu et al. [27] presented a morphing carbon fiber composite airfoil concept with an active trailing edge, which is enabled by an innovative structure driven by an electrical actuation system that uses linear ultrasonic motors (LUSM) with compliant runners.

One of the most effective methods of meeting the large-scale rapid change needs is to use deployable structures often called "structural mechanisms" [28], since they behave as mechanisms during the conversion process and resist loads when they are fixed [29,30]. The deployable structures have been widely surveyed and utilized in ordinary mechanical engineering [31], for example in complex space missions [32], small-scale structural applications [33], covering of swimming pools [34], bridge systems [35], and aerospace applications [36].

The most popular deployable structure is the scissor-structural mechanism (SSM), which is based on a scissor-like element (SLE), because they show effective performance by providing significant volume expansion, easy, and quick assembling/disassembling, requiring minimal damage to structural components during working [37]. With those advantages, for morphing wings, which require a large alteration of the skin and the body, the usage of SSMs can be a powerful solution.

Therefore, in this article, the synthesis, analysis, and design of SSMs, which consist of several types of SLEs for the trailing edge of a morphing wing, have been attempted, and it is shown that the

SSMs reduce the actuation mechanisms' complexity, produce good aerodynamic performance and require feasible torque.

2. Materials and Methods

This section details the theory behind the newly-developed actuation mechanism for the morphing of the trailing edge.

2.1. Scissor-Structural Mechanisms

2.1.1. Terms and Definitions

An SLE, also known as a pantograph or scissor-hinge unit, is a special type of kinematic element that is formed by two or more bars hinged together by a kinematic pair, often a revolute joint ("scissor-hinge") [38].

In Figure 1, the most basic form of an SLE is shown. In that, two straight bars are assembled to each other with a revolute joint. This location can be called the "pivot". In the same figure, end nodes, where an SLE can be assembled to another one, are shown as "hinge". The portion of a straight bar from the pivot to the hinges is called a "section". SLEs are distinguished by imaginary lines that go through the hinge locations, which are called "t-lines".

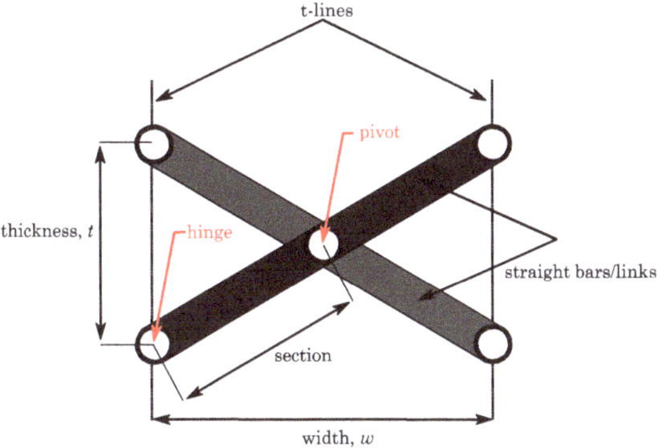

Figure 1. A basic scissor-like element

2.1.2. Typology of Scissor-Like Elements

SLEs have imaginary lines that go through the joint locations where an SLE can be assembled to another one (i.e., hinges) [39]. Calling those imaginary lines "t-lines", the group of "translational SLEs" covers the elements for which t-lines remain parallel throughout the deployment process. If the t-lines do not remain parallel throughout the deployment process, then this type of SLE is called "polar SLEs" or "curvilinear SLEs".

The difference between translational and polar SLEs can be revealed by a single quantity, which is the difference of the section ratios of bars from the SLEs. For example, let $l(k+1)$ and $l'(k'+1)$ be the lengths of the straight bars where k and k' determine the eccentricity of the hinge location. If $k = k'$, the SLE becomes a translational SLE; whereas, if $k \neq k'$, the SLE becomes a polar SLE. Figure 2 illustrates both translational and polar SLEs.

The common property of SLEs is the inverse-proportion of the thickness and the width. It is clearly seen that, by changing the type of SLE, such an inverse relation can take a complex form, which is used to stretch/shrink/bend any geometry in any direction. This relation can be brought out by defining a "foldability vector", which connects the midpoints of consecutive thicknesses. The detailed derivation of the foldability vector for each type of SLE can be found in [40,41].

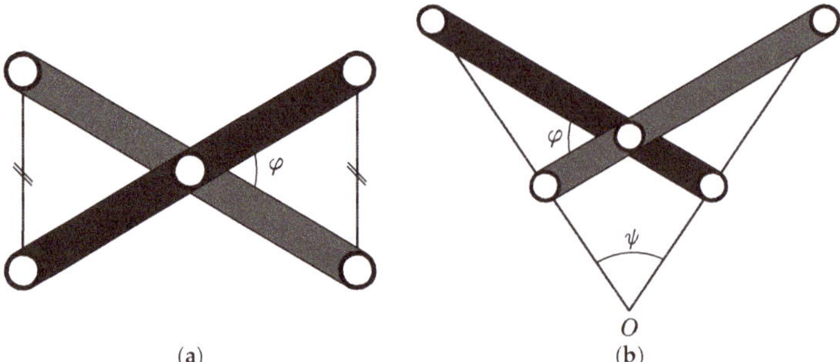

Figure 2. Two types of scissor-like element (SLE): (**a**) translational; (**b**) polar.

2.2. Kinematic Analysis of Scissor-Structural Mechanisms

In order to find out the kinematic capability of the designed mechanism where the resulting velocities and accelerations will be used in order to determine the inertial forces and compute the required torque to drive the mechanism, the kinematic analysis of a mechanism is required.

2.2.1. Mobility of Scissor-Structural Mechanisms

The mobility of an SSM can be defined as the number of input parameters (prime movers or actuators) that must be independently controlled to bring the mechanism into a desired position. It is also called as the degree-of-freedom (DOF) of a mechanism [42]. It is possible to calculate the mobility of the mechanism by directly counting the number of links and joints considering the types of mechanism, links, and joints through Chebyshev–Grübler–Kutzbach's formula. The Chebyshev–Grübler–Kutzbach's criterion is given as [43]:

$$M = \lambda\left(n_l - n_j - 1\right) + \sum_{i=1}^{n_j} f_i, \tag{1}$$

In Equation (1), λ represents the space in which the mechanism works. $\lambda = 3$ for planar and spherical mechanisms, while $\lambda = 6$ for spatial mechanisms. M is the DOF of the mechanism, n_l the total number of links, n_j the total number of joints, and f_i the independent DOF of the ith joint.

For the proposed planar SSM, which consists of N translational or polar SLEs of any kind, $\lambda = 3$, mobility can be calculated as: $M = 3(2N + 1 - 3N - 1) + 3N = 0$. If one of the fixed joints is set free, then mobility becomes $M = 2$.

According to these results, the proposed SSM cannot move (behaves as a structure) or needs two actuators to deploy; therefore, in order to obtain a single-DOF mechanism, one needs an additional input linkage system. Hence, an additional planar four-bar linkage (FBL), which was developed in [44], is assumed to be attached to manipulate the whole planar SSM.

2.2.2. Position Analysis of Scissor-Structural Mechanisms

Figure 3 represents the geometric representation of a generic planar SSM. In Figure 3, $C_j E_{j+1}$ and $E_j C_{j+1}$ are two straight bars that form the jth SLE of the SSM. l_j, $k_j l_j$ and l'_j, $k'_j l'_j$ represent section lengths, and γ_j and γ'_j represent the orientations about the global coordinates of those straight bars, respectively. When another SLE is attached to the jth SLE and letting the $(j + 1)$th SLE, there occurs a closed-loop $D_j C_{j+1} D_{j+1} E_{j+1}$, which is a "quadrilateral". This closed-loop can be expressed in complex notation as:

$$k_j l_j e^{i\gamma_j} + l'_{j+1} e^{i\gamma'_{j+1}} = k'_j l'_j e^{i\gamma'_j} + l_{j+1} e^{i\gamma_{j+1}}, \tag{2}$$

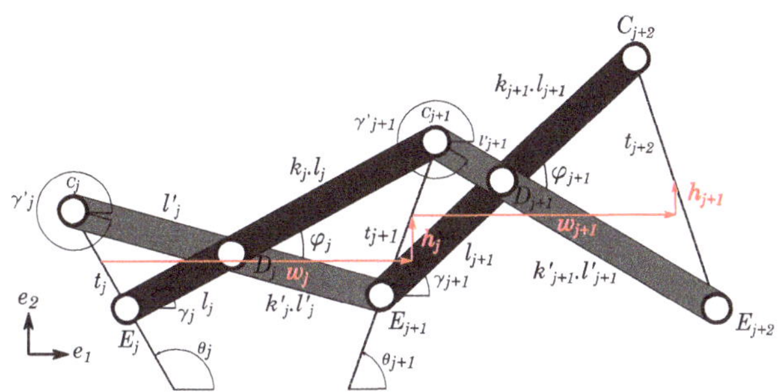

Figure 3. Geometric representation of a planar scissor-structural mechanism (SSM).

If the orientation angles of the jth SLE (γ_j and γ'_j) are known, Equation (2) can be solved. Assuming angle γ_j varies, then multiplying both sides of Equation (2) with $e^{-i\gamma'_j}$ gives an equation in terms of $\varphi_j \equiv \gamma_j - \gamma'_j$, $\mu_{j+1} \equiv \gamma_{j+1} - \gamma'_j$ and $\mu'_{j+1} \equiv \gamma'_{j+1} - \gamma'_j$:

$$k_j l_j e^{i\varphi_j} + l'_{j+1}e^{i\mu'_{j+1}} = k'_j l'_j + l_{j+1}e^{i\mu_{j+1}}, \tag{3}$$

In Equation (3), angle μ'_{j+1} can also be eliminated by using trigonometric identities. Hence, the resulting equation takes the form of the well-known Freudenstein equation when Euler's identity is applied:

$$F(\varphi_j, \mu_{j+1}) = p_1 + p_2\cos(\mu_{j+1}) - p_3\cos(\varphi_j) - \cos(\mu_{j+1} - \varphi_j) = 0, \tag{4}$$

where the Freudenstein parameters (p_1,p_2,p_3) are:

$$p_1 \equiv \frac{k'^2_j l'^2_j + k^2_j l^2_j - l'^2_{j+1} + l^2_{j+1}}{2k_j l_j l_{j+1}}, \; p_2 \equiv \frac{k'_j l'_j}{k_j l_j}, \; p_3 \equiv \frac{k'_j l'_j}{l_{j+1}}, \tag{5}$$

Equation (4) gives an implicit relation between the position variables φ_j and μ_{j+1}. This equation can be solved by applying half-tangent representation of the sine and cosine function in terms of unknown μ_{j+1}, which gives a quadratic equation in terms of $T \equiv \tan\left(\frac{\mu_{j+1}}{2}\right)$.

Since from the designed FBL, the first orientation angles γ_1 and γ'_1 are known, the other orientation angles, $\{\gamma_j\}_2^N$ and $\{\gamma'_j\}_2^N$ can be calculated through Equation (4). Hence, position vectors of all joints can be written in complex form as:

$$\begin{aligned}
\vec{r}_{C_{j+1}} &= \vec{r}_{E_j} + (l_j)(k_j+1)e^{i\gamma_j(t)}, \\
\vec{r}_{D_{j+1}} &= \vec{r}_{C_j} + l'_j e^{i\gamma'_j(t)}, \\
\vec{r}_{E_{j+1}} &= \vec{r}_{C_j} + (l'_j)(k'_j+1)e^{i\gamma'_j(t)},
\end{aligned} \tag{6}$$

The details of the theory and the extensive steps of the derivation can be found in [41].

2.2.3. Velocity and Acceleration Analysis of Scissor-Structural Mechanisms

In order to conduct a velocity and acceleration analysis, one first assumes that the angular velocity and acceleration of one of the links of the first SLE is prescribed (i.e., $\dot{\gamma}_1$, $\ddot{\gamma}_1$), then velocities and accelerations of the whole system can be calculated. The angular velocities and accelerations of the SSM can be determined by taking time derivatives of Equation (3) and writing the equations in matrix form [45].

Angular velocities can be determined as:

$$\{\dot{\mu}\} = [A]^{-1}\dot{\varphi}_j\{B_v\}, \tag{7}$$

where $[A]$ is known as the "characteristic matrix", $\{\dot{\mu}\}$ is the unknown angular velocity column matrix, and $\{B_v\}$ is the known column matrix. The details of these matrices are given in Equations (8)–(10).

$$[A] \equiv \begin{bmatrix} l_{j+1}\cos\mu_{j+1} & -l'_{j+1}\cos\mu'_{j+1} \\ l_{j+1}\sin\mu_{j+1} & -l'_{j+1}\sin\mu'_{j+1} \end{bmatrix}, \tag{8}$$

$$\{\dot{\mu}\} \equiv \left\{ \begin{array}{c} \dot{\mu}_{j+1} \\ \dot{\mu}'_{j+1} \end{array} \right\}, \tag{9}$$

$$\{B_v\} \equiv \left\{ \begin{array}{c} k_j l_j \cos\varphi_j \\ k_j l_j \sin\varphi_j \end{array} \right\}, \tag{10}$$

Since angular velocity $\dot{\gamma}_1$ is prescribed, and $\gamma = 0$; the other angular velocities, $\left\{\dot{\gamma}_j\right\}_2^N$ and $\left\{\dot{\gamma}'_j\right\}_2^N$, can be determined by solving Equation (7) for every SLE separately. Hence, the velocities of all joints can be written in complex form as:

$$\begin{aligned}
\vec{v}_{C_{j+1}} &= \vec{v}_{E_j} + i(l_j)(k_j+1)\dot{\gamma}_j(t)e^{i\gamma_j(t)}, \\
\vec{v}_{D_{j+1}} &= \vec{v}_{C_j} + il'_j\dot{\gamma}'_j(t)e^{i\gamma'_j(t)}, \\
\vec{v}_{E_{j+1}} &= \vec{v}_{C_j} + i\left(l'_j\right)\left(k'_j+1\right)\dot{\gamma}'_j(t)e^{i\gamma'_j(t)},
\end{aligned} \tag{11}$$

Similarly, the angular accelerations can be determined as:

$$\{\ddot{\mu}\} = [A]^{-1}\{B_a\}, \tag{12}$$

where the characteristic matrix $[A]$ is the same as Equation (8). $\{\ddot{\mu}\}$ is an unknown angular acceleration column matrix, and $\{B_a\}$ is the known column matrix, which are:

$$\{\ddot{\mu}\} \equiv \left\{ \begin{array}{c} \ddot{\mu}_{j+1} \\ \ddot{\mu}'_{j+1} \end{array} \right\}, \tag{13}$$

$$\{B_a\} \equiv \left\{ \begin{array}{c} k_j l_j \ddot{\varphi}_j \cos\varphi_j - k_j l_j \dot{\varphi}_j^2 \sin\varphi_j + l_{j+1}\dot{\mu}_{j+1}^2 \sin\mu_{j+1} - l'_{j+1}\dot{\mu}'^2_{j+1}\sin\mu'_{j+1} \\ k_j l_j \ddot{\varphi}_j \sin\varphi_j + k_j l_j \dot{\varphi}_j^2 \cos\varphi_j - l_{j+1}\dot{\mu}_{j+1}^2 \cos\mu_{j+1} + l'_{j+1}\dot{\mu}'^2_{j+1}\cos\mu'_{j+1} \end{array} \right\}, \tag{14}$$

Since angular acceleration $\ddot{\gamma}_1$ is prescribed, and $\ddot{\gamma}'_1 = 0$; other angular accelerations, $\left\{\ddot{\gamma}_j\right\}_2^N$ and $\left\{\ddot{\gamma}'_j\right\}_2^N$, can be determined by solving Equation (12) for every SLE separately. Hence, accelerations of all joints can be written in complex form as:

$$\begin{aligned}
\vec{a}_{C_{j+1}} &= \vec{a}_{E_j} - (l_j)(k_j+1)\left(\dot{\gamma}_j(t)\right)^2 e^{i\gamma_j(t)} + i(l_j)(k_j+1)\ddot{\gamma}_j(t)e^{i\gamma_j(t)}, \\
\vec{a}_{D_{j+1}} &= \vec{a}_{C_j} - l'_j\left(\dot{\gamma}'_j(t)\right)^2 e^{i\gamma'_j(t)} + il'_j\ddot{\gamma}'_j(t)e^{i\gamma'_j(t)}, \\
\vec{a}_{E_{j+1}} &= \vec{a}_{C_j} - \left(l'_j\right)\left(k'_j+1\right)\left(\dot{\gamma}'_j(t)\right)^2 e^{i\gamma'_j(t)} + i\left(l'_j\right)\left(k'_j+1\right)\ddot{\gamma}'_j(t)e^{i\gamma'_j(t)},
\end{aligned} \tag{15}$$

The details of the theory and the extensive steps of the derivation can be found in [41,46].

2.3. Aerodynamic Analysis of the Surfaces Formed by Scissor-Structural Mechanisms

In order to conduct an aerodynamic analysis for the surfaces formed by the SSM, the panel method is selected. It is widely known that the panel method provides solutions for linear, inviscid, irrotational flows around solid surfaces, which are subjected to subsonic speeds. The solution algorithm employs panels in order to construct the solid interfaces with vortices and sources inducing a velocity field around the body. The corresponding pressure distribution is calculated by using the tangential velocity components distributed over the surface, which also leads to the lift and drag forces by the integration of the pressure distribution over the airfoil contour. During the analysis, the computer program XFOIL (namely XFRL5) is utilized. Throughout the solutions performed using XFOIL, with $n_{panel} = 200$ panels assigned over the airfoil, the iteration number is limited to $n_{iter} = 100$, which is accompanied with a root-mean-square (RMS) tolerance of $tol = 10^{-4}$ for convergence. Angle-of-attack (AoA) values higher than $\alpha = 15°$ and lower than $\alpha = -5°$ were avoided. The detailed information about aerodynamic analysis can be found in [41,47].

2.4. Dynamic Force Analysis of Scissor-Structural Mechanisms

If inertial forces and moments are considered in the dynamic analysis of the mechanism, the analysis is called as the "dynamic force analysis". In order to determine the dynamic characteristics of the whole SSM, deriving the equations of one generic SLE is sufficient.

In the dynamic force analysis problem, in order to determine inertial forces and moments, arbitrary locations of the mass centers should be defined. The representation of arbitrary locations of mass centers of a generic SLE is given in Figure 4.

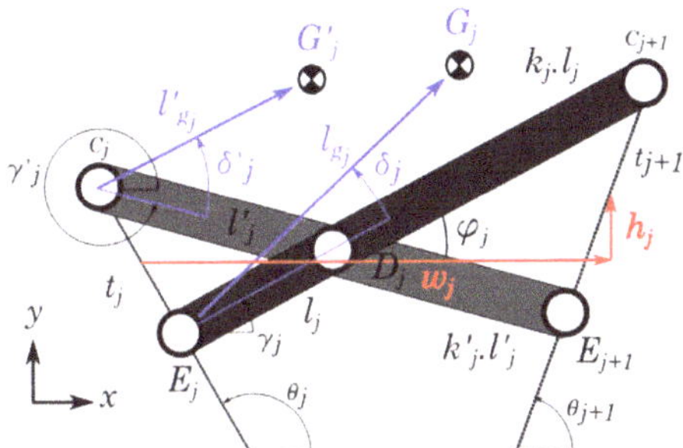

Figure 4. Representation of the arbitrary locations of mass centers of a generic SLE.

In Figure 4, G'_j and G_j represent the mass centers of links that construct an SLE; $l_{g'_j}$, l_{g_j} and δ'_j, δ_j represent the position vectors of the mass centers and the angles between the corresponding link and the mass center vector of that corresponding link, respectively.

Figure 5a shows the internal and external forces and moments on a generic SLE, while Figure 5b shows inertial forces and moments of the same element.

In order to determine the characteristic of the whole SSM, the system of equations represented in Equation (16) should be constructed and solved for every SLE separately. The equation system can be solved analytically with the information of SSMs having free ends; specifically, two hinges of the Nth SLE do not carry loads (i.e., $\vec{F}_{C_{N+1}} = \vec{F}_{E_{N+1}} = \vec{0}$). Therefore, internal forces on the other hinges can be calculated by solving Equation (16) starting from the Nth SLE to the first SLE (in the opposite direction of the kinematic analysis). Dynamic force analysis can be achieved by solving Equation (16) at different time steps. The details of the theory and the extensive steps of the derivation can be found in [41,48].

$$\{f\} = \left[A_f\right]^{-1}\{B_f\}, \tag{16}$$

where $\{f\}$ is the column matrix, which contains unmown internal forces:

$$\{f\} = \left\{ \ \left(F_{C_j}\right)_x \ \ \left(F_{D_j}\right)_x \ \ \left(F_{E_j}\right)_x \ \ \left(F_{C_j}\right)_y \ \ \left(F_{D_j}\right)_y \ \ \left(F_{E_j}\right)_y \ \right\}^T, \tag{17}$$

Using trigonometric identities, the matrix $\left[A_f\right]$ can be constituted as:

$$\left[A_f\right] = \begin{bmatrix} 1 & -1 & 0 & 0 & 0 & 0 \\ 0 & 1 & 1 & 0 & 0 & 0 \\ 0 & 0 & 0 & 1 & -1 & 0 \\ 0 & 0 & 0 & 0 & 1 & 1 \\ l'_{gj}\mathrm{s}\left(\gamma'_j+\delta'_j\right) & -g_{D'_j}\mathrm{s}(\eta_{D'j}) & 0 & -l'_{gj}\mathrm{c}\left(\gamma'_j+\delta'_j\right) & g_{D'_j}\mathrm{c}(\eta_{D'j}) & 0 \\ 0 & g_{D_j}\mathrm{s}(\eta_{Dj}) & l_{gj}\mathrm{s}(\gamma_j+\delta_j) & 0 & -g_{D_j}\mathrm{c}(\eta_{Dj}) & -l_{gj}\mathrm{c}(\gamma_j+\delta_j) \end{bmatrix}, \tag{18}$$

where $\mathrm{s}(x) \equiv \sin(x)$ and $\mathrm{c}(x) \equiv \cos(x)$.

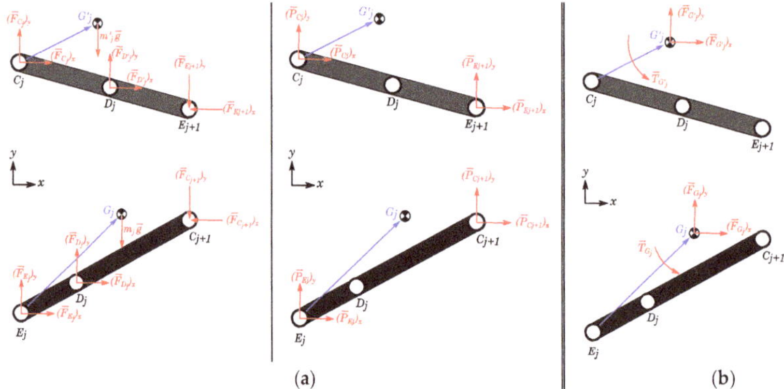

Figure 5. Free-body-diagram of an SLE: (**a**) internal-external forces, (**b**) effective forces.

The known column matrix $\left\{B_f\right\}$ can be constituted as:

$$\left\{B_f\right\} = \left\{ \begin{array}{c} \left(F_{G_j}\right)_x + \left(F_{E_{j+1}}\right)_x - \left(P_{C_j}\right)_x - \left(P_{E_{j+1}}\right)_x \\ \left(F_{G_j}\right)_x + \left(F_{C_{j+1}}\right)_x - \left(P_{E_j}\right)_x - \left(P_{C_{j+1}}\right)_x \\ \left(F_{G_j}\right)_y + \left(F_{E_{j+1}}\right)_y - \left(P_{C_j}\right)_y - \left(P_{E_{j+1}}\right)_y + m'_j g \\ \left(F_{G_j}\right)_y + \left(F_{C_{j+1}}\right)_y - \left(P_{E_j}\right)_y - \left(P_{C_{j+1}}\right)_y + m_j g \\ T_{G'j} - l'_{gj}\left(P_{C_j}\right)_x \mathrm{s}\left(\gamma'_j+\delta'_j\right) + l'_{gj}\left(P_{C_j}\right)_y \mathrm{c}\left(\gamma'_j+\delta'_j\right) + g_{E_{j+1}}\left(\left(F_{E_{j+1}}\right)_x - \left(P_{E_{j+1}}\right)_x\right)\mathrm{s}(\eta_{Ej+1}) - g_{E_{j+1}}\left(\left(F_{E_{j+1}}\right)_y - \left(P_{E_{j+1}}\right)_y\right)\mathrm{c}(\eta_{Ej+1}) \\ T_{Gj} - l_{gj}\left(P_{E_j}\right)_x \mathrm{s}(\gamma_j+\delta_j) + l_{gj}\left(P_{E_j}\right)_y \mathrm{c}(\gamma_j+\delta_j) + g_{C_{j+1}}\left(\left(F_{C_{j+1}}\right)_x - \left(P_{C_{j+1}}\right)_x\right)\mathrm{s}(\eta_{Cj+1}) - g_{C_{j+1}}\left(\left(F_{C_{j+1}}\right)_y - \left(P_{C_{j+1}}\right)_y\right)\mathrm{c}(\eta_{Cj+1}) \end{array} \right\}, \tag{19}$$

In Equations (18) and (19), $\vec{g}_{D_j}$, $\vec{g}_{C_{j+1}}$, $\vec{g}_{D'_j}$, $\vec{g}_{E_{j+1}}$ represent vectors from the mass centers to the hinges and the pivot. These vectors and their orientations can be defined as:

$$\begin{aligned} \vec{g}_{D_j} &\equiv l_{gj}e^{i(\gamma_j+\delta_j)} - l_j e^{i\gamma_j} \ \text{and} \ \eta_{Dj} \equiv \arg \vec{g}_{D_j}, \\ \vec{g}_{C_{j+1}} &\equiv l_{gj}e^{i(\gamma_j+\delta_j)} - l_j(k_j+1)e^{i\gamma_j} \ \text{and} \ \eta_{Cj+1} \equiv \arg \vec{g}_{C_{j+1}}, \\ \vec{g}_{D'_j} &\equiv l'_{gj}e^{i(\gamma'_j+\delta'_j)} - l'_j e^{i\gamma'_j} \ \text{and} \ \eta_{D'j} \equiv \arg \vec{g}_{D'_j}, \\ \vec{g}_{E_{j+1}} &\equiv l'_{gj}e^{i(\gamma'_j+\delta'_j)} - l'_j\left(k'_j+1\right)e^{i\gamma'_j} \ \text{and} \ \eta_{Ej+1} \equiv \arg \vec{g}_{E_{j+1}}, \end{aligned} \tag{20}$$

After the calculation of all internal forces and moments, the required driving torque of the SSM can be determined accordingly. Equation (19) considers the internal forces and externally-applied pressure forces, as well. Hence, it is suitable for the case where the aerodynamic effects are also taken into the consideration. In Equation (19), F's denote the internal forces, and P's stand for the external forces due to the pressure; therefore, the same equation can also be used for the dynamic force analysis of the SSM in vacuo by simply assigning P's equal to zero.

2.5. Synthesis and Design of Scissor-Structural Mechanisms

Any planar SSM constructed by using the proposed translational and polar SLEs can generate three different 2D curves. As can be seen from Figure 6, those three curves are assumed to pass through joint locations $\{C_j\}_1^N$, $\{D_j\}_1^N$, $\{E_j\}_1^N$, respectively. This property of planar SSMs can be used to morph structures that are described by curves.

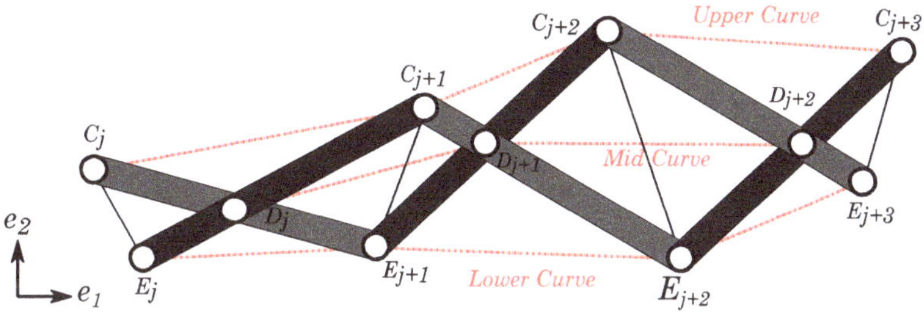

Figure 6. Curves that can be generated by a planar SSM.

It is convenient to start the design with the selection of the total number of used SLEs (N). By segmentation, "baseline curves", which characterize the original structure, and "target curves", which is the morphed form of the structure, can be analyzed to find the correct number of used SLE.

With the power of the computer, the above design procedure can be expedited by perturbing the design parameters. The obtained SSMs can be compared by their errors, and the final SSM that best matches both the baseline and target curves can be selected [41,46]. Such a type of error, which originates in the design, is named "design error" and can be defined as:

$$err_j = \frac{d_j}{s},\tag{21}$$

where in Equation (21), d_j is the shortest distance from the newest joint locations of jth SLE to the target curves and s is the characteristic dimension selected to reveal the percentage error.

Since all errors are lengths, so positive $\left(\{err_j\}_1^N \geq 0\right)$, the mean error can be calculated as:

$$err_m = \frac{1}{N}\sum_1^N err_j,\tag{22}$$

which is the only parameter to compare different candidate SSMs with each other and select the best one, which has the minimum mean design error.

In this study, the theoretical parts explained so far are written in METU-licensed MATLAB software [49]. The flowchart of the complete developed computer-routine is given in Appendix A.

3. Results

This section presents the various results of the study.

3.1. Scissor-Structural Mechanism for the Trailing Edge of a Morphing Wing

In the study, the chord length of the airfoil was taken as 0.6 m, and the rear spar of the wing was taken at 60% of the chord length, similar to [41,50]. In this paper, the actuator, which was an FBL and was intended to drive the whole SSM, was located in the torque box of the UAV wing. The FBL was assumed to be attached to the SSM and drive the SSM from the first SLE that was the closest one to the rear spar.

In Figure 7a, the SSM with $N = 6$ SLEs is considered. The baseline airfoil was NACA 4412, and the target airfoil was NACA 8412. In this case, mean t-line orientation angle was determined as a hundred and ten degrees, $\overline{\theta} = 110°$. All SLEs were chosen as the type of polar-SLE with constant $\{\psi_j\}_1^N = 4°$. Segmentation was done linearly for simplicity (the width of each SLE was equal). Then, when the anchor-link was rotated $\Delta\phi = 22.5°$ in the clockwise direction, the designed SSM generated the NACA 8412 geometry with 0.11% mean design error.

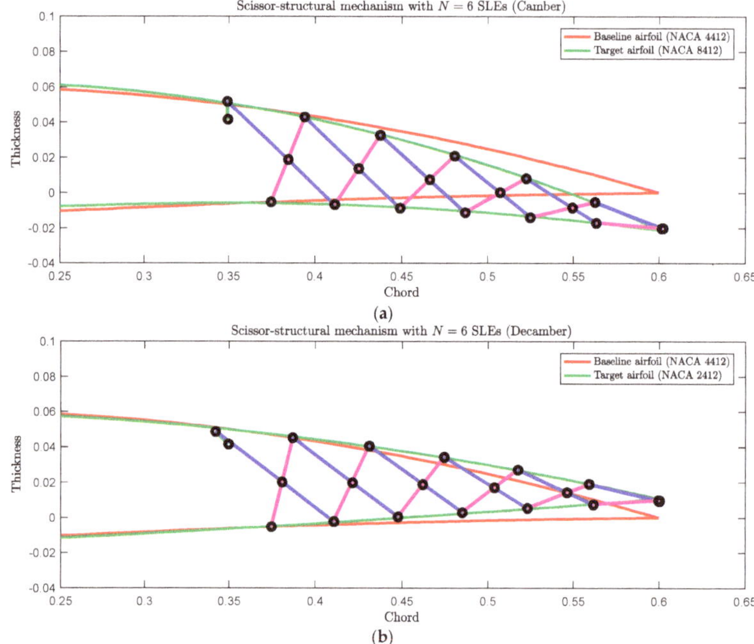

Figure 7. Scissor-structural mechanism with N = 6 SLEs at the deployed position when the target airfoil was: (**a**) NACA 8412; (**b**) NACA 2412.

As seen from Figure 7b, if the anchor-link was rotated in the counter-clockwise sense from the baseline airfoil of NACA 4412, the mechanism also added the decamber property to the aircraft wing. The same SSM with $N = 6$ SLEs could eventually generate the NACA 2412 profile with 0.125% mean design error.

This study assumed that the wing skin was composed of a fully-compliant material, which followed any prescribed motion during the morphing. Such a wing concept, which was enhanced by the SSM, required the segmentation or discretization of the wing skin in the chordwise direction. Each wing skin segment corresponded to a particular SLE and changed its position and orientation together with the movement of that SLE. Hence, each skin segment could be considered as comprised of two different portions. The hinges of SLEs were assumed to be attached to the wing skin from the rigid portion of the skin segment, whereas those rigid portions were combined together with a hyper-elastic portion. Therefore, the wing skin could be formed as a sequence of rigid and hyper-elastic portions successively.

This study only considered the development of an internal actuation mechanism. However, in order to predict the characteristic of the required surface material, displacements of the wing segments which were in between two consecutive SLEs were also considered.

Figure 8a gives the first pose of the SSM with $N = 6$ SLEs, which stretched the wing skin or produced elongation with a magnitude of at most 3.5% when it morphed the wing profile into NACA 8412; whereas, Figure 8b gives the second pose of the SSM, which stretched the wing skin with a magnitude of at most 1.5% when it morphed the wing profile into NACA 2412.

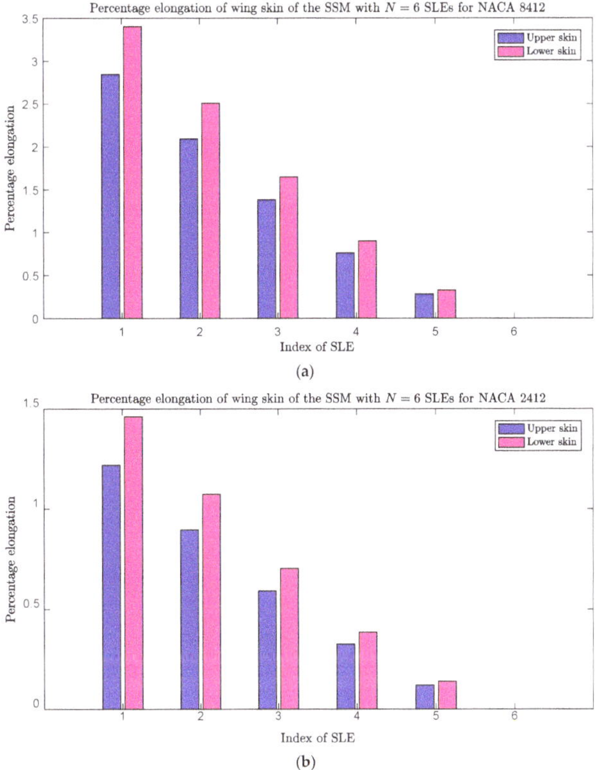

Figure 8. Percentage elongation of wing skin segments of the SSM with N = 6 SLEs for: (**a**) NACA 8412; (**b**) NACA 2412.

SSMs can affect the wing skin by stretching or shrinking the hyper-elastic portion of the wing skin segment. In order to avoid any slack of the hyper-elastic portion, the whole wing skin should be in tension. Moreover, due to structural concerns, such extensions should be small enough. As seen from the results, if the SSM was allowed to manipulate the chord length, the designed SSM stretched the wing skin up to 3.5%. Therefore, the slackness of the wing skin will be avoided by using SSMs within feasible limits, which is an advantageous property. Another result was that the designed SSM required a compliant material for the elastic portion of the wing skin, which provided a safe elongation up to 3.5%. Such an extension capability corresponds to materials in the literature that are good enough for aircraft wings.

3.2. Aerodynamic Analysis of the Surfaces Formed by Scissor-Structural Mechanisms for the Trailing Edge of a Morphing Wing

Aerodynamic analyses were conducted for the designed SSMs with the package XFLR5, which is an analysis tool for airfoils, wings, and planes operating at low Reynolds numbers. XFLR5 includes XFOIL's direct and inverse analysis capabilities with wing design and analysis capabilities based on the lifting line theory, on the vortex lattice method, and on a three-dimensional (3D) panel method [51].

Figure 9 shows the aerodynamic behavior of the surfaces formed by the SSM with $N = 6$ SLEs when the SSM morphed the baseline wing profile into the position of NACA 8412 and NACA 2412, respectively. It can be seen that the surfaces formed by the SSM nearly show as similar performance as the NACA airfoils.

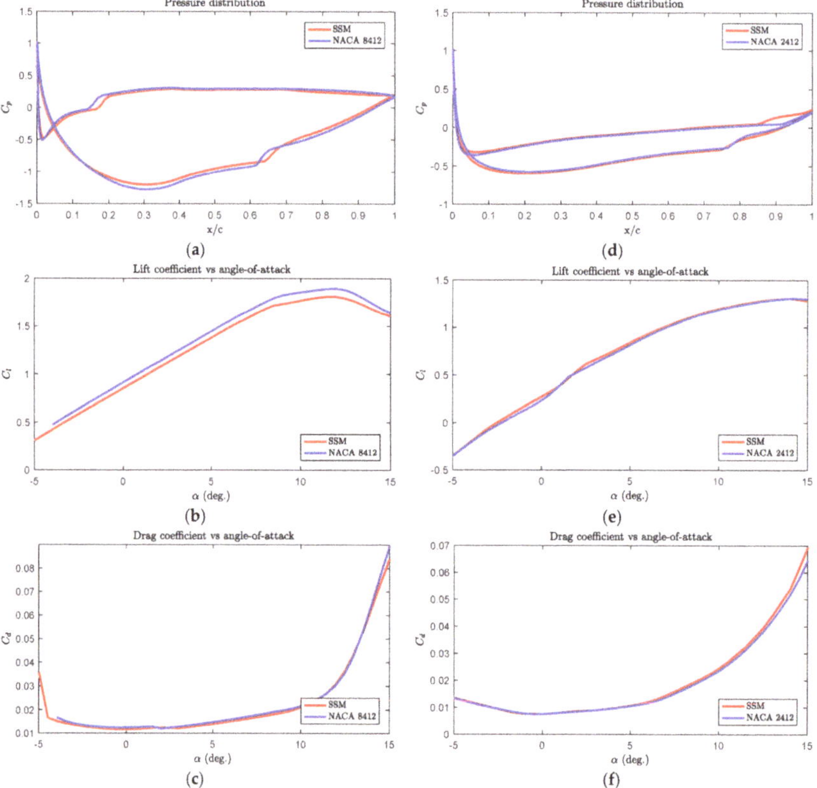

Figure 9. Comparison of the aerodynamic behavior of the surface formed by the proposed SSM with $N = 6$ SLEs at the position of the baseline airfoil morphed into NACA 8412: (**a**) pressure distribution; (**b**) lift coefficient vs. AoA; (**c**) drag coefficient vs. AoA; and the surface formed by the proposed SSM with $N = 6$ SLEs at the position of the baseline airfoil morphed into NACA 2412: (**d**) pressure distribution; (**e**) lift coefficient vs. AoA; (**f**) drag coefficient.

3.3. Dynamic Force Analysis of the Scissor-Structural Mechanisms for the Trailing Edge of a Morphing Wing

Dynamic force analysis of the mechanism can be performed in two conditions, which are "in in vacuo" and "under aerodynamic loading".

First of all, one should define masses and mass centers. For that purpose, the masses of links were calculated by assuming the material of the links: aluminum with $\rho_{Al} = 2700 \left(\frac{\text{kg}}{\text{m}^3} \right)$. The volumes of the links and their moment of inertias were calculated assuming that the links were those of rectangular beams with a square cross-section. Both sides of each unique element were one eighth (1/8) of the airfoil thickness. The mass centers of each element were assumed to be located at their geometric centers. By considering these parameters, the weight penalty of the mechanism brought to the wing was 138.7 g, where the weight of the aircraft is 25 (kg).

3.3.1. In Vacuo Condition

In Figure 10a, the magnitudes of internal forces of selected links (links of the first, the middle, and the last SLEs) of the SSM with $N = 6$ SLEs while it was forming the baseline airfoil into NACA 8412 are shown.

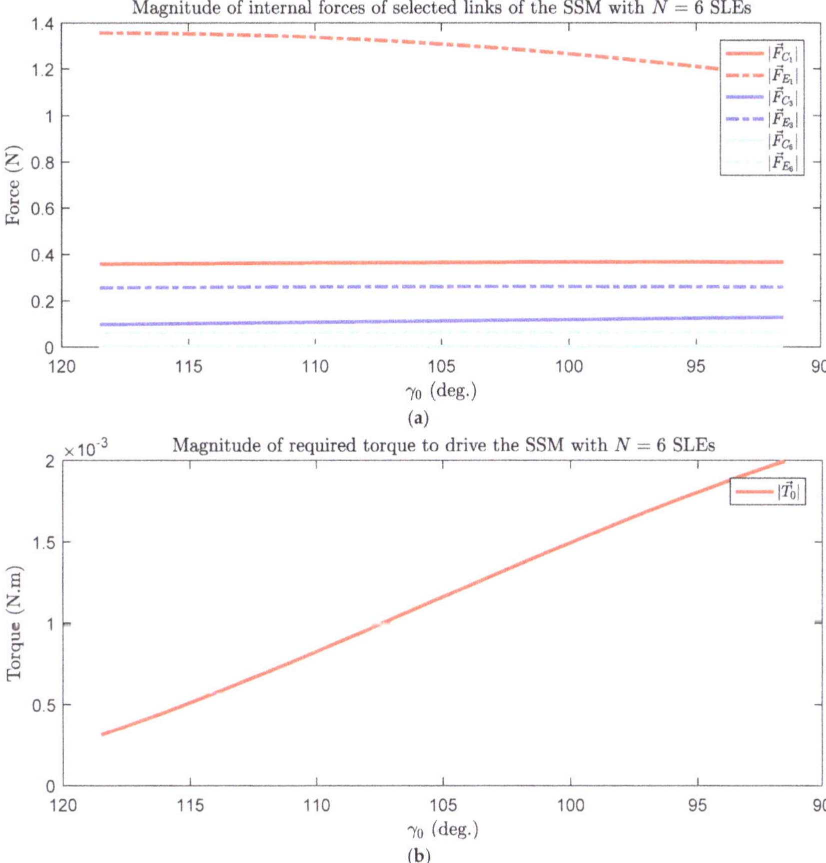

Figure 10. (a) Magnitude of internal forces of selected links of the SSM and (b) Magnitude of required torque to drive the SSM with $N = 6$ SLEs for NACA 8412.

In Figure 10b, the magnitude of required torque to drive the SSM with $N = 6$ SLEs while it was forming the baseline airfoil into NACA 8412 is shown.

It can be seen from Figure 10 that only those links that were close to the torque application carried some internal forces due to the applied torque, whereas the other links carried almost zero internal forces. The main characteristic of such a type of deployable structure is having a stress-strain free state. The computed very low internal forces and moments are important in the verification of this fact.

Moreover, the required torque variation in in vacuo condition was below 0.002 (N·m), which is very low since the SSM is too light and aerodynamic loading is ignored. This torque value can be associated with the dynamic characteristics of the analysis, which can be neglected if the aerodynamic forces and moments are considered.

3.3.2. Under Aerodynamic Loading

The calculated pressure coefficient distribution of the surface formed by the SSM could be used to estimate the required torque to drive the SSM under aerodynamic loading. In order to convert the pressure distribution, C_p, into the nodal forces on the upper and lower surface hinge locations of the SSM, the sea level properties of the air were used. The air velocity was assumed as 0.2 Mach. For the simplicity of the problem, a single C_p distribution was used for all poses of the SSM.

In Figure 11, the magnitudes of the internal forces of the selected links and the magnitude of the required torque of the SSM with $N = 6$ SLEs in order to morph the baseline airfoil into NACA 8412 are shown, respectively.

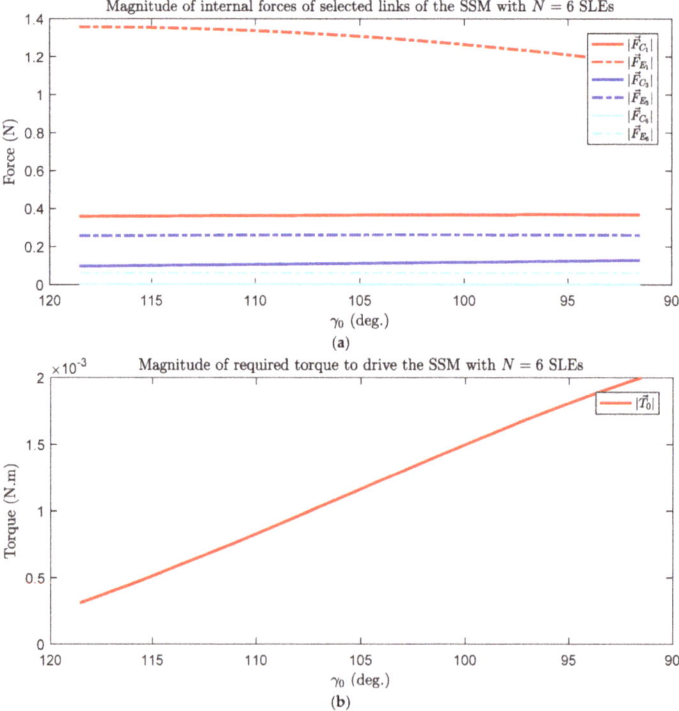

Figure 11. (a) Magnitude of internal forces of selected links of the SSM and (b) magnitude of the required torque to drive the SSM with $N = 6$ SLEs for NACA 8412.

As can be seen from Figures 10 and 11, the aerodynamic loading increased the computed internal forces and the torque values. However, the maximum torque value still remained within the capabilities of UAV torque motors, which varied up to 0.4 (N·m) [50].

4. Discussion

In this study, the synthesis, analysis, and design of a special type of deployable mechanism, which is the scissor-structural mechanism (SSM), for morphing of the trailing edge of an aircraft wing is presented. The wing skin is assumed to be a hybrid one with fully-compliant and rigid segments, and in order to satisfy the mobility requirements, a four-bar linkage (FBL) is assumed to be attached to the designed SSM.

A computer-routine, which synthesizes, analyzes, and design an SSM in order to morph different NACA airfoils into different shapes is developed. A single sample SSM with six SLEs is presented. The results show that the developed mechanism satisfies both camber and decamber target profiles with mean design errors of 11% and 125%, respectively.

The profiles obtained from the proposed mechanism are modelled and analyzed aerodynamically with the XFLR5 (panel method). The obtained results are compared with NACA airfoils. The results show that surfaces formed by SSMs for each case nearly produce the same pressure distribution, lift, and drag as the target NACA airfoils.

The dynamic force analysis of the designed SSM has also been performed in order to compute the internal force values that occur in the elements of the mechanism and the required torque value that is necessary for driving the whole SSM. The computed results revealed that as the links get farther from the application point, the force that carries it progressively gets lower. That shows that the link which is farthest away carries almost zero internal force. It is also computed that the required torque value is very small in in vacuo condition. However, aerodynamic loading was found to have increased the required torque value significantly. Therefore, it is believed that the aerodynamic loading is critical for the determination of the size of the actuator and the links.

Author Contributions: Conceptualization Y.Y.; methodology, H.L.Ş. and Y.Y.; software, H.L.Ş.; writing H.L.Ş. and, Y.Y.; visualization, H.L.Ş.; supervision, Y.Y.

Funding: This research received no external funding.

Conflicts of Interest: The authors declare no conflict of interest.

Appendix A

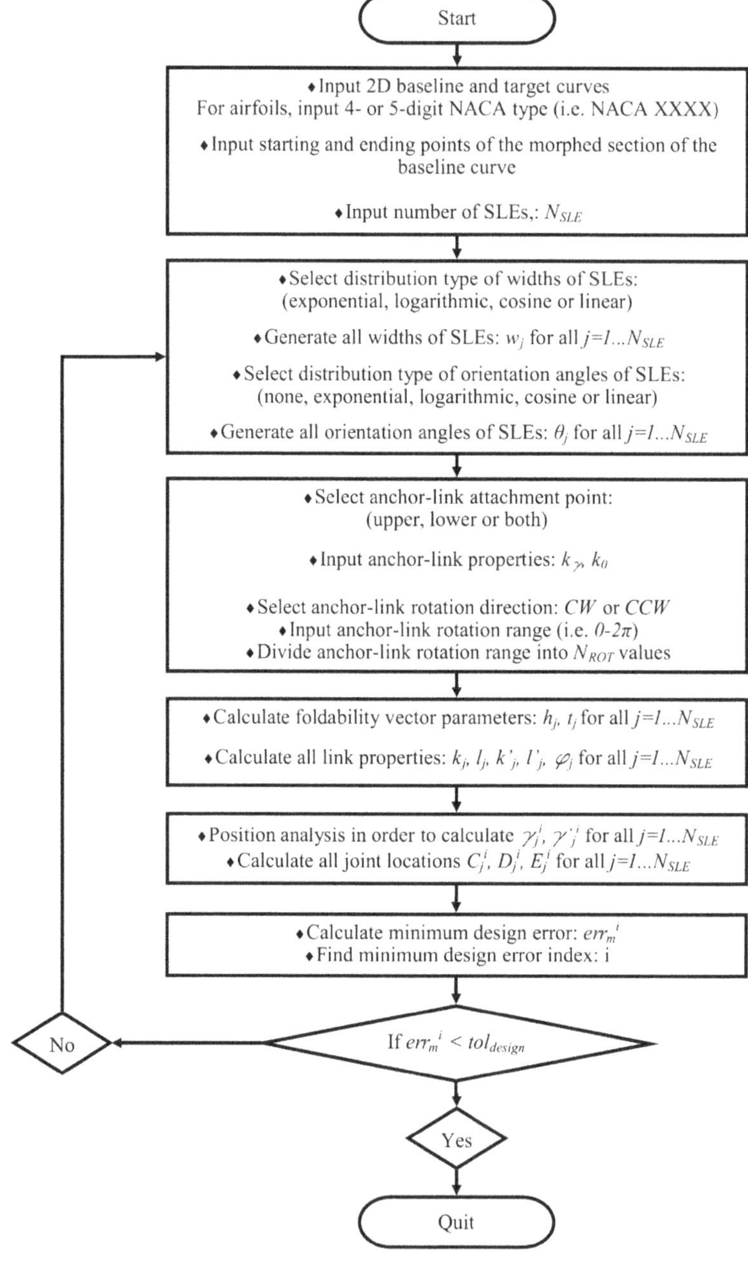

Figure A1. Flowchart of the computer routine developed, which synthesizes the SSM for given baseline and target curves.

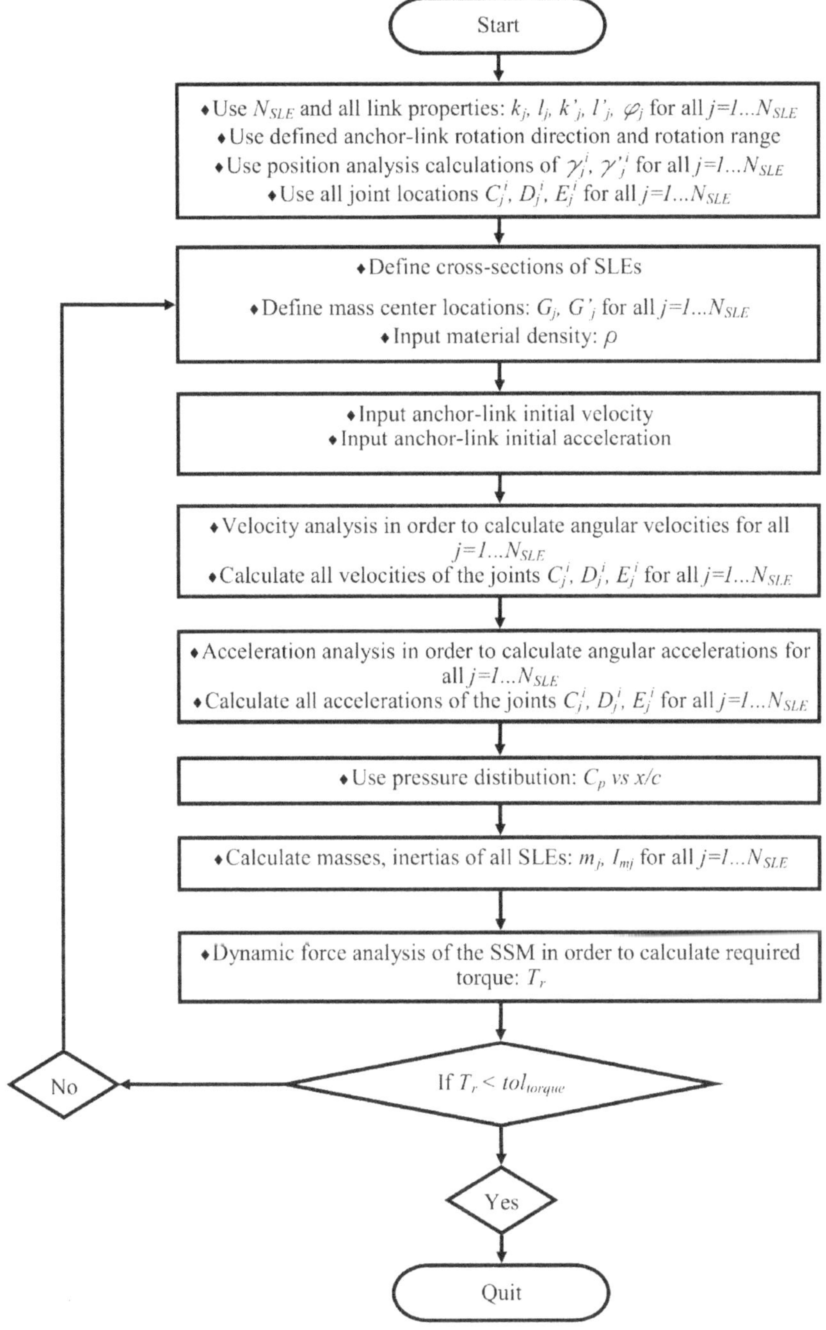

Figure A2. Flowchart of the computer routine developed, which analyzes the synthesized SSM.

References

1. Ghommem, M.; Hajj, M.R.; Beran, P.S.; Puri, I.K. Role of Wing Morphing in Thrust Generation. *Theor. Appl. Mech.* **2014**, *4*, 1–7. [CrossRef]
2. Barbarino, S.; Bilgen, O.; Ajaj, R.M.; Friswell, M.I.; Inman, D.J. A Review of Morphing Aircraft. *J. Intell. Mater. Syst. Struct.* **2011**, *22*, 823–877. [CrossRef]
3. Ninian, D.; Dakka, M.S. Design, Development and Testing of Shape Shifting Wing Model. *Aerospace* **2017**, *4*, 52. [CrossRef]
4. Fincham, J.H.S.; Friswell, M.I. Aerodynamic Optimisation of a Camber Morphing Aerofoil. *Aerosp. Sci. Technol.* **2015**, *43*, 245–255. [CrossRef]
5. Olympio, K.R.; Gandhi, F. Flexible Skins for Morphing Aircraft Using Cellular Honeycomb Cores. *J. Intell. Mater. Syst. Struct.* **2010**, *21*, 1719–1735. [CrossRef]
6. Kudva, J.N. Overview of the DARPA Smart Wing Project. *J. Intell. Mater. Syst. Struct.* **2004**, *15*, 261–267. [CrossRef]
7. Thornton, S.V. *Reduction of Structural Loads Using Maneuver Load Control on the Advanced Fighter Technology Integration (AFTI)/F-111 Mission Adaptive Wing*; National Aeronautics and Space Administration (NASA): Washington, DC, USA, 1993.
8. Smith, K.; Butt, J.; von Spakovsky, M.; Moorhouse, D. A Study of the Benefits of Using Morphing Wing Technology in Fighter Aircraft Systems. In Proceedings of the 39th AIAA Thermophysics Conference, Fluid Dynamics and Co-Located Conferences, Miami, FL, USA, 25–28 June 2007; American Institute of Aeronautics and Astronautics (AIAA): Miami, FL, USA, 2007.
9. Sofla, A.Y.N.; Meguid, S.A.; Tan, K.T.; Yeo, W.K. Shape Morphing of Aircraft Wing: Status and Challenges. *Mater. Des.* **2010**, *31*, 1284–1292. [CrossRef]
10. Abbott, I.H.; Von Doenhoff, A.E. *Theory of Wing Sections: Including a Summary of Airfoil Data*, 2nd ed.; Dover Publications Inc.: New York, NY, USA, 1959; Volume 11, ISBN 0486605868.
11. Raymer, D.P. *Aircraft Design: A Conceptual Approach*, 4th ed.; American Institute of Aeronautics and Astronautics, Inc.: Washington, DC, USA, 2006; ISBN 1563478293.
12. Murugan, S.; Woods, B.K.S.; Friswell, M.I. Hierarchical Modeling and Optimization of Camber Morphing Airfoil. *Aerosp. Sci. Technol.* **2015**, *42*, 31–38. [CrossRef]
13. Vasista, S.; Tong, L.; Wong, K.C. Realization of Morphing Wings: A Multidisciplinary Challenge. *J. Aircr.* **2012**, *49*, 11–28. [CrossRef]
14. Parker, H.F. Variable-Camber Rib for Aeroplane-Wings. US Patent No. 1341758, 1 June 1920.
15. Bonney, L.W. Aeroplane Wing or Aerofoil Structure. US Patent No. 1710673, 23 April 1929.
16. Antoni, U. Construction of Flexible Aeroplane Wings Having a Variable Profile. US Patent No. 1886362, 8 November 1932.
17. Grant, C.H. Airfoil. US Patent No. 2022806, 3 December 1935.
18. Cole, J.B. Variable Camber Airfoil. US Patent No. 4053124, 11 October 1977.
19. Zapel, E.J. Variable Camber Trailing Edge for Airfoil. US Patent No. 4131253, 26 December 1978.
20. Monner, H.P.; Breitbach, E.; Bein, T.; Hanselka, H. Design Aspects of the Adaptive Wing—The Elastic Trailing Edge and the Local Spoiler Bump. *Aeronaut. J.* **2000**, *104*, 89–95. [CrossRef]
21. Bartley-Cho, J.D.; Wang, D.P.; Martin, C.A.; Kudva, J.N.; West, M.N. Development of High-rate, Adaptive Trailing Edge Control Surface for the Smart Wing Phase 2 Wind Tunnel Model. *J. Intell. Mater. Syst. Struct.* **2004**, *15*, 279–291. [CrossRef]
22. Campanile, L.F.; Anders, S. Aerodynamic and Aeroelastic Amplification in Adaptive Belt-Rib Airfoils. *Aerosp. Sci. Technol.* **2005**, *9*, 55–63. [CrossRef]
23. Marques, M.; Gamboa, P.; Andrade, E. Design of a Variable Camber Flap for Minimum Drag and Improved Energy Efficiency. In Proceedings of the 50th AIAA/ASME/ASCE/AHS/ASC Structures, Structural Dynamics, and Materials Conference, Structures, Structural Dynamics, and Materials and Co-Located Conferences, Palm Springs, CA, USA, 4–7 May 2009; American Institute of Aeronautics and Astronautics (AIAA): Palm Springs, CA, USA, 2009.

24. Barrett, R.V.; Roelof, V. Mechanics of Pressure-Adaptive Honeycomb and Its Application to Wing Morphing. *Smart Mater. Struct.* **2011**, *20*, 94010.

25. Takahashi, H.; Yokozeki, T.; Hirano, Y. Development of Variable Camber Wing with Morphing Leading and Trailing Sections Using Corrugated Structures. *J. Intell. Mater. Syst. Struct.* **2016**, *27*, 2827–2836. [CrossRef]

26. Pecora, R.; Magnifico, M.; Amoroso, F.; Monaco, E. Multi-Parametric Flutter Analysis of a Morphing Wing Trailing Edge. *Aeronaut. J.* **2014**, *118*, 1063–1078. [CrossRef]

27. Wu, R.; Soutis, C.; Zhong, S.; Filippone, A. A Morphing Aerofoil with Highly Controllable Aerodynamic Performance. *Aeronaut. J.* **2017**, *121*, 54–72. [CrossRef]

28. Akgün, Y.; Gantes, C.J.; Sobek, W.; Korkmaz, K.; Kalochairetis, K. A Novel Adaptive Spatial Scissor-Hinge Structural Mechanism for Convertible Roofs. *Eng. Struct.* **2011**, *33*, 1365–1376. [CrossRef]

29. Chen, Y. Design of Structural Mechanisms. Ph.D. Thesis, University of Oxford, Oxford, UK, 2003.

30. Şahin, H.L.; Yaman, Y. Design and Analysis of a Mechanism for the Chord and Camber Morphing of an Aircraft Wing. In Proceedings of the 7th EASN International Conference on Innovation in European Aeronautics Research, Warsaw, Poland, 26–28 September 2017; The European Aeronautics Science Network (EASN): Warsaw, Poland, 2017.

31. Zhao, J.S.; Chu, F.; Feng, Z.J. The Mechanism Theory and Application of Deployable Structures Based on SLE. *Mech. Mach. Theory* **2009**, *44*, 324–335. [CrossRef]

32. Chen, W.-J.; Luo, Y.-Z.; Fu, G.-Y.; Gong, J.-H.; Dong, S.-L. A Study on Space Masts Based on Octahedral T russ Family. *Int. J. Space Struct.* **2001**, *16*, 75–82. [CrossRef]

33. Rosenfeld, Y.; Logcher, R.D. New Concepts for Deployable-Collapsable Structures. *Int. J. Space Struct.* **1988**, *3*, 20–32. [CrossRef]

34. Escrig, F.; Perez Valcarcel, J.; Sanchez, J. Deployable Cover on a Swimming Pool in Seville. *J. Int. Assoc. Shell Spat. Struct.* **1996**, *37*, 39–70.

35. Thrall, A.P.; Adriaenssens, S.; Paya-Zaforteza, I.; Zoli, T.P. Linkage-based movable bridges: Design methodology and three novel forms. *Eng. Struct.* **2012**, *37*, 214–223. [CrossRef]

36. You, Z.; Pellegrino, S. Foldable Bar Structures. *Int. J. Solids Struct.* **1997**, *34*, 1825–1847. [CrossRef]

37. Hanaor, A.; Levy, R. Evaluation of Deployable Structures for Space Enclosures. *Int. J. Space Struct.* **2001**. [CrossRef]

38. Akgün, Y. A Novel Transformation Model for Deployable Scissor-Hinge Structures. Ph.D. Thesis, İzmir Institute of Technology, İzmir, Turkey, 2010.

39. Roovers, K.; De Temmerman, N. Deployable Scissor Grids Consisting of Translational Units. *Int. J. Solids Struct.* **2017**, *121*, 45–61. [CrossRef]

40. Langbecker, T. Kinematic Analysis of Deployable Scissor Structures. *Int. J. Space Struct.* **1999**, *14*, 1–15. [CrossRef]

41. Şahin, H.L. Synthesis, Analysis and Design of a Novel Mechanism for the Trailing Edge of a Morphing Wing. Master's Thesis, Middle East Technical University (METU), Ankara, Turkey, 2018.

42. Uicker, J.J., Jr.; Pennock, G.R.; Shigley, J.E. *Theory of Machines and Mechanisms*, 3rd ed., Oxford University Press, Inc.: New York, NY, USA, 2003; ISBN 019515598X.

43. Zhao, J.; Feng, Z.; Ma, N.; Chu, F. *Design of Special Planar Linkages*, 1st ed.; Springer: Berlin/Heidelberg, Germany, 2014; ISBN 978-3-662-51353-8.

44. Şahin, H.L.; Yaman, Y. Design of a Mechanism that Provides Camber and Chord Change for a Fully Morphing Wing. In Proceedings of the 6th National Conference on Aerospace, Kocaeli, Turkey, 28–30 September 2016; UHUK: Kocaeli, Turkey, 2016. (In Turkish)

45. Russell, K.; Shen, Q.; Sodhi, R.S. *Mechanism Design: Visual and Programmable Approaches*, 1st ed.; CRC Press, Inc.: Boca Raton, FL, USA, 2013; ISBN 1466570172.

46. Şahin, H.L.; Yaman, Y. Design and Analysis of a Novel Mechanism for the Morphing of Trailing Edge of an Aircraft Wing. *MATEC Web Conf.* **2018**, *188*, 04001. [CrossRef]

47. Şahin, H.L.; Çakır, B.O.; Yaman, Y. Aerodynamic Modelling and Analysis of a Novel Mechanism for Chord and Camber Morphing Wing. *MATEC Web Conf.* **2018**, *188*, 04002. [CrossRef]

48. Şahin, H.L.; Çakır, B.O.; Yaman, Y. Dynamic Force Analysis of a Novel Mechanism for Chord and Camber Morphing Wing Under Aerodynamic Loading. *MATEC Web Conf.* **2018**, *233*, 00006. [CrossRef]

49. MATLAB®. Available online: https://www.mathworks.com/products/matlab.html (accessed on 9 November 2018).
50. Spirlet, G.B. Design of Morphing Leading and Trailing Edge Surfaces for Camber and Twist Control. Master's Thesis, Delft University of Technology, Delft, The Netherlands, 2015.
51. XFLR5. Available online: http://www.xflr5.com/xflr5.htm (accessed on 9 November 2018).

aerospace

Review

Innovative Scaled Test Platform e-Genius-Mod—Scaling Methods and Systems Design

Dominique Paul Bergmann *, Jan Denzel, Asmus Baden, Lucas Kugler and Andreas Strohmayer 

Institute of Aircraft Design (IFB), University of Stuttgart, Pfaffenwaldring 31, 70569 Stuttgart, Germany; denzel@ifb.uni-stuttgart.de (J.D.); st102967@stud.uni-stuttgart.de (A.B.); st146408@stud.uni-stuttgart.de (L.K.); strohmayer@ifb.uni-stuttgart.de (A.S.)

* Correspondence: bergmann@ifb.uni-stuttgart.de; Tel.: +49-711-685-60342

Received: 30 November 2018; Accepted: 8 February 2019; Published: 14 February 2019

Abstract: Future aircraft design highly depends on the successful implementation of new technologies. However, the gap between conventional designs and new visions often comes with a high financial risk. This significantly complicates the integration of innovations. Scaled unmanned aircraft systems (UAS) are an innovative and cost-effective way to get new configurations and technologies in-flight. Therefore the Institute of Aircraft Design developed the e-Genius-Mod taking into account all relevant similitude requirements. It is a scale model of the electric motor glider e-Genius. Since the Reynolds number for the free-flight model cannot be adhered to, an airfoil was developed with lift-to-drag and lift-to-angle-of-attack courses reproducing the full-scale e-Genius flight characteristics. This will enable testing and assessment of new aviation technologies in a scaled version with an opportunity for free-flight demonstration in relevant environment.

Keywords: unmanned aircraft; scaling; aircraft design; flight testing; technology readiness level; technology demonstrator

---

1. Introduction

This article is an extension of the conference paper "UAS as flexible and innovative test platform for aircraft configuration and systems testing" [1].

For many years unmanned aircraft systems (UAS) have played an important role in aviation research. UAS were used especially for development and testing of new aircraft configurations in different research projects. For example, the configuration of a blended wing body for commercial aviation was successfully tested with a scaled unmanned free-flying platform. The collected flight data gave first insights into flight behavior [2,3] as well as the impact on passenger comfort of such an unknown configuration in flight. Beside Computational Fluid Dynamics (CFD) analysis and wind tunnel tests, UAS free flight models can be a cost-efficient alternative for the investigation of new aircraft configurations, as shown in various research projects.

1.1. Future Challenges

Beside classic approaches to aircraft design, new requirements and technologies have to be considered in future designs to meet imminent political and economic requirements. "Flightpath 2050—Europe's Vision for Aviation" defines specific challenges for future aircraft [4]. One important design driver is to protect the environment, with requirements to:

- Reduce CO_2 and NO_x emissions
- Reduce noise emission

- Taxi emission free

To meet these requirements, new technologies are required in aviation, such as for example electric or hybrid-electric powertrains. Especially in the field of emission reduction, these technologies can offer an advantage over conventional propulsion systems. However, the efficient and profitable use of these propulsion systems also requires an adaptation of the aircraft configuration. This technology is already partially available in innovative research concepts, which must be brought to a demonstration in flight. Once tested and proven in flight the technology will have a bigger impact than on a laboratory test level. These technology changes pose a high risk when it comes to truly redeveloping and certifying such systems for commercial use. A look at the development costs for new aircraft reveals a significant increase [4] in development costs per seat. This cost increase is aggravated when new technologies are used, such as for the A350 or B787 [5]. This is one of the main reasons why todays aircraft configurations typically rely on well-known configuration types and technologies. Commercialization of novel technologies therefore typically takes a long time from the actual invention to market entry.

To be able to meet future challenges as quickly and efficiently as possible and to minimize the financial risk for the manufacturers, the challenge will be to offer an economically viable way to bring innovative technologies to an appropriate technology readiness level (TRL), where they get interesting for commercialization due to a significantly lower technological and financial risk. In addition to risk reduction, development time also has to be reduced. Today the development of new technologies (flying components) for aviation takes around 20 years from TRL1 to TRL9 [5].

1.2. Accelerate Technology Readiness

Testing aircraft models in wind tunnels is an established technique of validating new prototype designs. An alternative is the testing with free-flight models based on a UAS test platform. This form of investigating flight characteristics, using scaled models, is mainly used in research projects, with a focus on the investigation of the flight dynamics of novel unconventional aircraft configurations.

Over and above the investigation of new aircraft configurations, UAS can be used as testbeds for new aircraft technologies. The process to bring new technologies or research concepts for aviation to flight is time and cost intensive. This fact proves to be a major hurdle to a fast and innovative technology transfer, which allows many innovative ideas to be realized only to an initial TRL of 1–3.

The proposed method to use UAS as test and evaluation platforms should facilitate an economical and cost-effective aircraft design process and the commercialization of innovative technologies.

To accelerate the TRL of new technologies and concepts, UAS can become a handsome tool, promoting an aircraft design which will be innovative and competitive on the international market due to novel concepts and technologies. UAS should not replace manned flight testing. However, they can be an extending tool, providing a link between upstream research and technology demonstration. In this way, UAS as test beds for technology investigation and demonstration can be an additional tool besides classic analytic design methods and CFD calculation. As UAS flight campaigns can be less restrictive than manned flight testing, provided there is a restricted test area, new technologies can be integrated faster and tested earlier in flight. This fast technology demonstration in relevant environment brings significant benefits in the development process. Not only can concept issues be identified in early design phases, reducing cost for changes, but also potential regulatory barriers can be identified and properly addressed at an early stage. This allows reducing the financial risk for commercializing projects of innovative technologies by identifying and eliminating potential development risks. Feasibility and impact of technologies can be demonstrated at early development stages, making them more attractive and predictable to bring them from science to business.

2. Technology Test Bed e-Genius-Mod

In order to investigate the benefits of UAS as testbeds to demonstrate and evaluate new aviation technologies and aircraft configurations, the project e-Genius-Mod was started at the Institute of Aircraft Design (IFB) at University of Stuttgart. The project is targeted at demonstrating new technologies on a scaled testbed in flight to accelerate the development of new aircraft technologies.

A scaled model as well as a full-scale aircraft is required to close the complete investigation loop. At this point the question has to be solved which kind of aircraft configuration is a meaningful basis for the investigation of this process in an academic research environment. Scaled models of previous IFB projects targeted on passenger aircraft configurations [6]. Classical aircraft configurations [7] are an interesting field to investigate new technologies, especially under commercial aspects. For an academic project like e-Genius-Mod this would be of interest, but it would be challenging to prove the research concepts, as no full scale aircraft is directly accessible, in order to compare and evaluate the results of a scaled flight demonstration campaign. The e-Genius is ideal to understand the similarity conditions between scale and full-scale aircraft, as IFB has full access to it, and therefore can make modifications and carry out full-scale flight tests. Also the e-Genius database built up over years of flight testing is available for validating the scaled test results.

E-Genius is a two seat full electric motor glider. The full-scale e-Genius is a manned technology test aircraft. E-Genius in Figure 1 is built to demonstrate the competitiveness of electrical propulsion over conventional propulsion systems in aviation. It has a wingspan of 16.9 m and a maximum take-off mass of 950 kg. The electric power train allows a maximum range of more than 400 km.

Figure 1. e-Genius (University of Stuttgart, Institute of Aircraft Design).

The model e-Genius-Mod is prepared as a technology test bed to demonstrate new technologies for future aircraft design in relevant environment. The UAS test bed will be used for academic and innovative research projects to demonstrate new technologies up to TRL6. The design of the test bed is initially prepared for the investigation of new aircraft configuration solutions for distributed electric propulsion systems.

Once the similarity requirements between the scaled model and the full scale aircraft are identified, the model is a pre-test bed for the full-scale aircraft. Results like improved performance can be collected and the most promising concept can be identified for full-scale integration.

The free-flight platform e-Genius-Mod (Figure 2) is a 33.3% scale model of the e-Genius, scaled in compliance with the similarity conditions of Froude number and Reynolds number as a free flight model [3]. The model is prepared as a technology test bed to demonstrate new technologies for future aircraft design in relevant environment. The UAS test bed will be used for academic and innovative research projects to investigate scaling similarities of free flight models and to demonstrate new technologies up to TRL6. The design of the test bed is primarily prepared for the investigation of new aircraft configuration solutions for distributed electric propulsion systems.

The test bed e-Genius-Mod will be an extension of the full-scale aircraft for further investigation of electrical flight. Electrical propulsion systems provide new degrees of freedom for aircraft design with respect to propulsion integration. To investigate and demonstrate these innovations, a flexible test bed like the unmanned scale model is required. The demonstration of innovative concepts like distributed propulsion on a manned aircraft is expensive and time intensive and not suitable for research with unknown outcomes. With its modular airframe design the e-Genius-Mod is the basis for an efficient and systematic research of the various effects of distributed propulsion.

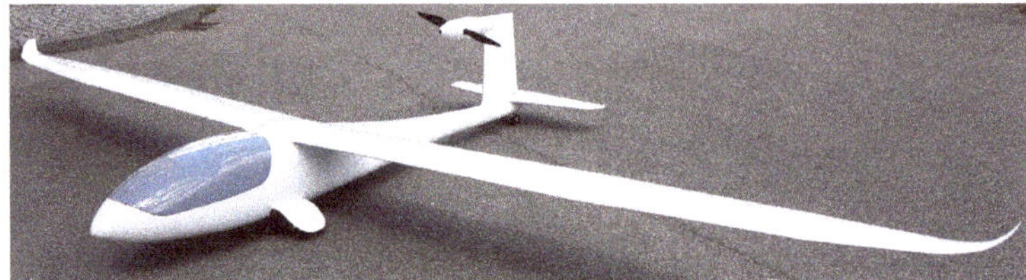

Figure 2. e-Genius-Mod 33.3% free-flight model.

2.1. Scaling

To realize an efficient test bed, scaling relationships and similarity requirements have to be adhered to [8,9]. To get reliable flight data from the free flight test with a model aircraft the following aspects must be respected for scaling:

- Dimensions
- Aerodynamics
- Inertia

Scaling of the e-Genius-Mod followed the scientific findings of NASA Technical Paper 1435 "Similitude Requirements and Scaling Relationships as Applied to Model Testing" [3]. Dimension and inertia scaling follow the similarity criteria of Froude number [3,10].

$$Fr_{scale} = Fr_{full-scale} \tag{1}$$

Aspects of aerodynamic scaling must mainly respect the Reynolds number. With respect of the conditions for similarity of Froude number the following equation for the Reynolds number is fulfilled [3]:

$$Re_{scale} = Re_{full-scale} \cdot \left(\frac{\nu_{full-scale}}{\nu_{scale}} \right) \cdot n^{\frac{3}{2}} \tag{2}$$

Airfoil of the wing and aerodynamic surfaces has to be proven with respect to their effectiveness while scaling under Froude number.

2.1.1. Scaling Factor

The scaling factor is a key factor for the usefulness of the UAS test bed. Several aspects must be considered for the identification of the matching scale factor:

- Flight conditions and test area
- Handling and transportability
- Environmental influences
- Payload test equipment
- Scalability of new technologies which shall be investigated

Furthermore, a big advantage of the e-Genius is its dimension of a general aviation aircraft, which allows a small scaling factor. It is assumed that the error influence increases with increasing scaling factor. The larger this error is in the overall scaling of all the variables relevant for flight test, the lower the transferability of results from a scale model to its full-scale equivalent is estimated. By determining the scale factor, previous research results must be taken into account, which show that similarity of Reynolds number cannot be met for free-flight models [9]. Based on this knowledge, when considering the Reynolds number of the model, it was taken into account that it does not lie within the critical range of Reynolds numbers for the selected scaling factor. To obtain the longest possible attached air flow, the Reynolds number is in the supercritical range. For the e-Genius-Mod the following scaling factor and Reynolds number are chosen [11]:

$$n = 1:3\ (33.3\%)\ Re_{scale} = 467948\ \text{(related to wing root)}$$

The Reynolds number of the model of 467.948 (operating altitude 300 m) has to be seen alongside a Reynolds number for the full-scale aircraft of 2057088 (operating altitude 2000 m) [11]. This means an error of 22.7% for the Reynolds number must be taken into account in further designs.

2.1.2. Airfoil Scaling

As opposed to the geometric scale of the full-scale aircraft, the airfoil of the main wing cannot be scaled geometrically without influencing the flight characteristics as the Reynolds number cannot be met. To ensure the model airplane to have the same flight properties, the fundamental dimensionless quantities of the main wing should not have too large of deviations [12]. Otherwise flight performance tests of the model cannot deliver valid results for investigating scaling laws as a technical basis comparing the full-scale and the scale model of the e-Genius and other airplanes.

The most important dimensionless quantity describing the properties of an airfoil of the wing is the Reynolds number. It characterizes the flow formation in the boundary layer. Due to the differences in flight altitude of the scale model, speed and geometric scale of the depth of the sash airfoil the Reynolds number cannot be met. This manifests itself by different transition spot positions, where laminar flow switches to turbulent.

The unavoidable deviations of some dimensionless quantities such as the Reynolds number result in massive deviations comparing the polar curves from full-scale and scaled aircraft. The main goal is to ensure the same flight characteristics by creating adherence of the polar curves from the full-scale and scaled aircraft [11,13]. An effective approach to this problem is made by doing some geometrical modifications of the airfoil. The following parameters are modified:

- Thickness/chord ratio
- Airfoil camber
- Location of maximum thickness
- Location of maximum chamber
- Leading edge radius

To reduce the complexity of the wing with its different segments the wing panel is reduced to an airfoil with constant chord, constant thickness and no twist. The primary focus is on the flight performance which is characterized by the flight condition achieving maximum range. To align the polar curves of scaled and full-scale airfoils the angle of attack for the flight at maximum range has to be determined.

The given design velocity of the e-Genius is 43.056 m/s [11]. The scaled velocity can be calculated under similarity of the Froude number for free-flying models [3,9], as the flight conditions of full-scale and scaled e-Genius are always under incompressible flow. "For incompressible flow, the kinematic properties are preserved by using velocities scaled from Froude number similitude requirements (Froude scaling)." [3] (p. 2).

$$v_{scale} = v_{full-scale} \cdot n^{\frac{1}{2}} = 24.858 \frac{m}{s} \tag{3}$$

Equation (3) leads into a scaled velocity of the scale model of 24.858 m/s. Substituting the velocity into the lift equations in horizontal flight leads to the lift coefficient. Marking the lift coefficient in the polar curves of the airfoil calculated using X-foil as aerodynamic tool supplies the desired value for the angle of attack ($\alpha = 3.5°$). The following development of the airfoil of the scaled aircraft is made, regarding the best possible conformity in the area near to an angle of attack of 3.5°, as further investigations focus on cruise flight conditions.

Regarding the polar curve of both the scaled and the full-scale e-Genius, the laminar area of the curve of the scaled model is way more differentiated and has wider dimensions. In general the curve is displaced towards higher drag coefficients. Both on the upper and on the lower side of the airfoil the transition from laminar to turbulent flow in the boundary layer takes place later towards the trailing edge. The moment coefficients of scaled and full-scale airfoil are nearly similar, the value of the model being slightly smaller [11].

To correct the form of the laminar area, the thickness/chord ratio is reduced significantly [13]. The earlier occurring suction peak at the upper boundary of airfoil of the wing displaces the transition spot towards the leading edge and reduces the width of the laminar area of the polar curve.

The airfoil camber is reduced to reduce the slightly increased level of the lift coefficient, which also moves the laminar area in direction of the laminar section of the full-scale. The movement of the location of maximum chamber towards the leading edge moves the spot of transition in the same direction. It is important to check the effect of any geometrical change on the angle of attack where no lift occurs. Any changes in that angle result in a change of the velocity of maximum range.

To compensate for the negative effect of a thickness reduction of the airfoil with respect to stalling behavior, the location of maximum thickness is moved towards the leading edge. The risk of uncontrolled flow separation in the boundary layer is reduced by the homogenization of the pressure coefficient in the rear of the airfoil. In addition, the level of the drag coefficient is reduced in the turbulent areas of the polar curve and helps to assimilate the curves of full-scale and scaled e-Genius-Mod.

The polar curves of the model are adapted to those of the original by stepwise manual adjustment of the modifications listed above.

The installation of tabulators delivers good results to optimize the length of laminar flow sections regarding the prioritized flight condition of maximum range. They ensure the transition from laminar to turbulent flow in the boundary layer exactly is in the right place creating a flow formation around the airfoil very similar to that of the original aircraft. However, the losses of laminar flow areas at different angles of attack have to be taken into account creating higher drag in other flight conditions.

To ensure sufficiently good flight characteristics beside the flight performance tests, the following flight conditions require closer examination [11–13]:

- Stall at high angles of attack (Figure 3)
- Flight characteristics at low speed/landing (Figure 4)
- Ensure maneuverability

The main issue regarding high angles of attack is the decreasing lift coefficient beyond the laminar areas of the polar curve and the early stall of the original full-scale airfoil. Reducing thickness and camber as well as moving the location of maximum thickness towards the leading edge aligns the curves of the scaled and the full-scale aircraft up to an angle of attack of $16°$.

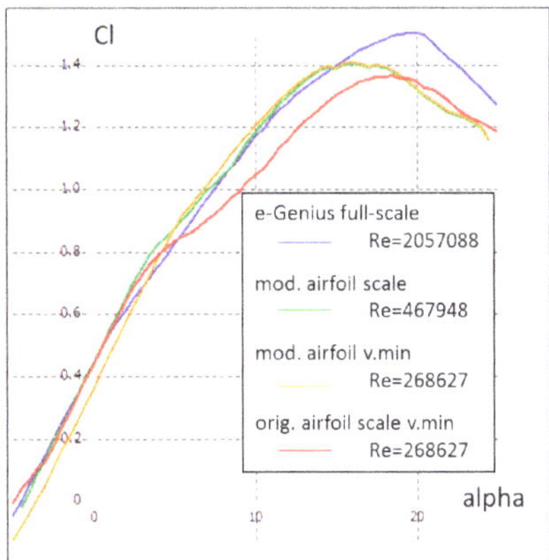

Figure 3. Lift to angle of attack—full-scale vs. scale airfoil [11].

The required lift coefficients calculated using the lift equation are being compared with those from the polar curves to verify the ability of holding altitude at the scaled minimum velocity using the modified airfoil. Although the required lift cannot be created, the polar curves of the modified airfoil align much better to those of the full scale aircraft compared to the original one. The occurring error in the achieved landing velocity is 19% [11].

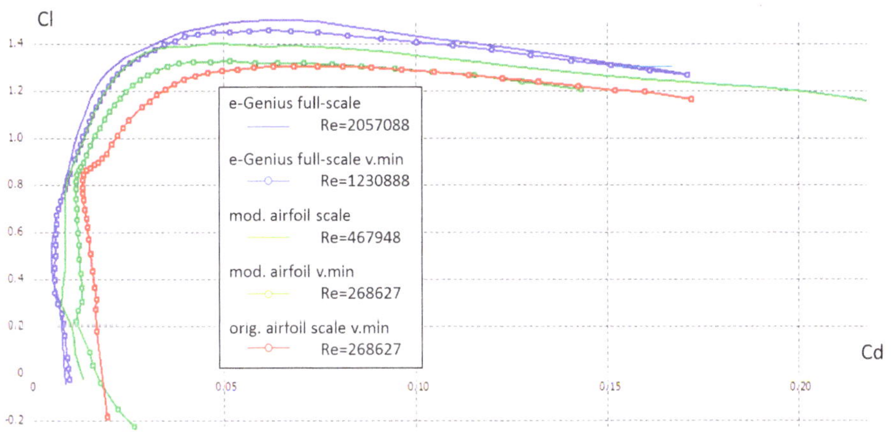

Figure 4. Lift to drag—full-scale vs. scale airfoil [11].

The curve progression of the moment coefficient at different states of surface actuators deflection is examined to ensure the maneuverability (Figure 5) of the scaled aircraft. All in all the responses in all flight conditions are nearly identical, only at very high angles of attack and low speed the effects of the surface actuators are apparently smaller.

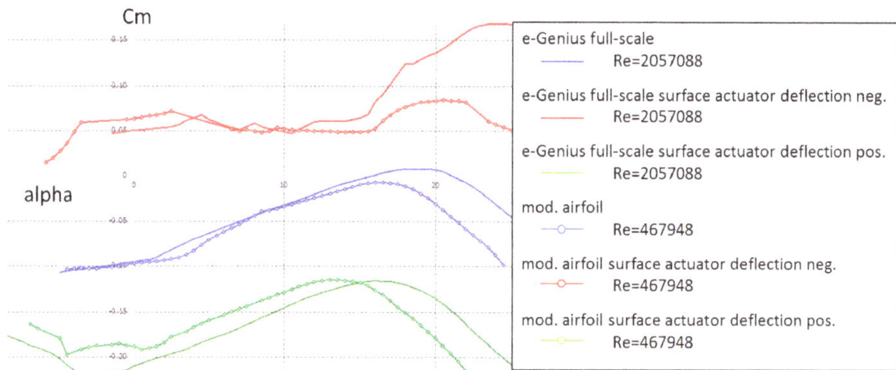

Figure 5. Insurance of the maneuverability - full-scale vs. scale airfoil [11].

Figure 6 shows the full-scale and the scaled airfoil which allows inflight investigations at cruise speed with similarity requirements. The technical data of the airfoils are shown in Table 1.

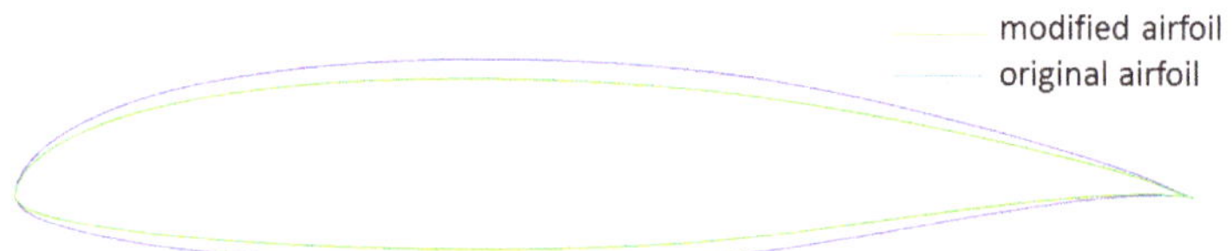

Figure 6. Geometrical change of the airfoil [11].

Table 1. Airfoil data full-scale vs. scale [11].

Profil Parameter	Original Full-Scale Airfoil	Modified Scale Airfoil
Thickness/chord ratio (%)	17.01	14.00
Airfoil camber (%)	2.71	2.62
Location of maximum thickness (%)	40.40	39.39
Location of maximum camber (%)	39.39	38.38
Leading-edge radius (m)	0.02026	0.01383

2.1.3. Variable Pitch Propeller

In order to model several flight conditions in a realistic way with respect to the propulsion system, the scaled e-Genius-Mod is fitted with a constant speed propeller system. Hereby it is possible to operate the model in its optimal operating points, for example during climb or cruise (Figure 7). A variable pitch propeller allows operating the propulsion components in the point of their maximum point of efficiency.

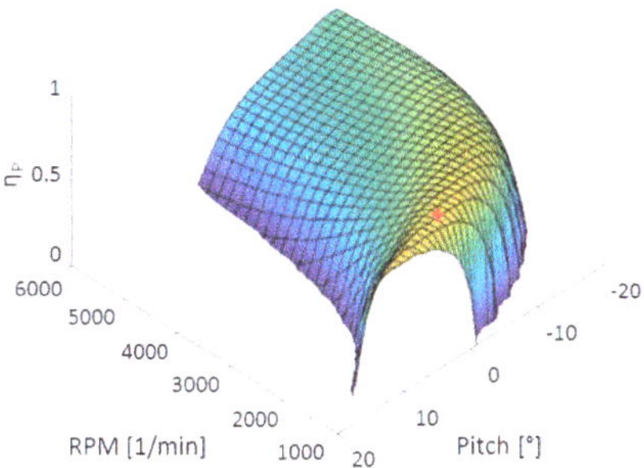

Figure 7. Propeller efficiency at 17 m/s (true air speed) [14].

To stay close to the original, a variable pitch propeller (VPP) has also been developed specifically for the scaled model [14]. Due to the relative rarity of variable pitch propellers for UAS of this size, it was decided to conduct a study on the e-Genius-Mod propeller system to compare it with the fixed propeller, and to determine the optimal settings for different flight situations. The overall goal is to gather information to optimize the operation of the propeller system.

For this purpose, a simulation program based on blade-element theory was set up in Matlab [14]. To obtain the geometrical data needed for the calculations in a nondestructive way, an image-based process is used [15] to extract the angle of incidence and chord as functions of the radius. The program could then be used to compute the propeller performance as a function of flight speed pitch and RPM, and plot this data in 2D or 3D charts. Furthermore, an optimization algorithm is implemented to find the best settings in different situations.

As expected, the program allowed to get a good insight in the performance that can be expected from the e-Genius powertrain and the resulting flight performance, for the variable-pitch as well as the fixed-pitch propeller. The results permitted to draw very interesting conclusions:

- The fixed propeller can efficiently cover a large portion of the flight envelope. In cruise flight (best range or best endurance), the fixed propeller works very close to its optimum, and the VPP barely gives any performance improvement (less than 2%). This is due to the e-Genius being a 33.3% scaled model, and having a smaller flight envelope than that of a full-scale aircraft.
- During climbing flight, the VPP improves performance by approximately 25%. However, the e-Genius-Mod, being a UAS, will mostly fly at low altitude (<500m). Consequently, climbing will be very short when compared to the cruise flight, thus limiting the performance improvement.
- The electric engine has a very good efficiency over a wide range of RPM, therefore the advantage of running the engine near its point of best efficiency (the way it would be done on a piston-engine aircraft) does not translate to the e-Genius model.
- The pitch variation system itself weighs 400 g, and due to its position at the aft of the fuselage this weight needs to be counterbalanced in the nose to keep a neutral center of gravity. Consequently, the overall weight penalty ranges from 400 to 1200 g (depending if the payload itself can be used as ballast to counterbalance this without adding overall weight, but this may not always be possible). Therefore, in most flight situations the increased weight of the variable-pitch propeller might negate the possible performance improvements.

In certain situations however, it could be useful to install a VPP:

- If the runway is particularly short, take-off performance can be noticeably improved.
- If the flight tasks require flying at either very low or very high speeds, outside the range where the fixed propeller performs well.

Therefore the decision to install the system or not should be taken after careful analysis of the future flight task and operating conditions. The results (Figure 8) highlight how the intended flight task of an aircraft, as well as the characteristics of its powertrain influences the usefulness of a variable-pitch propeller.

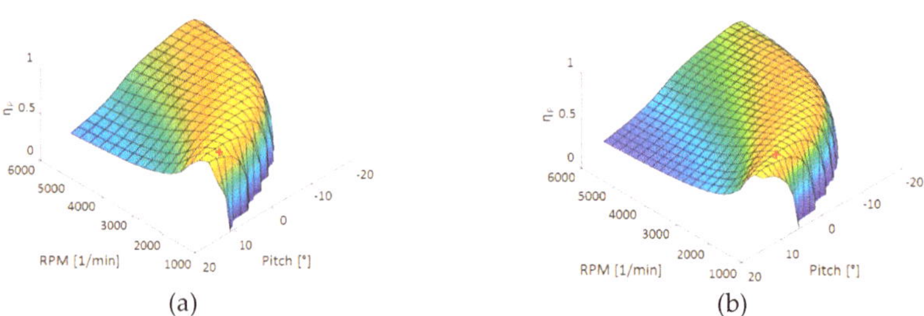

(a)   (b)

Figure 8. VVP for max range (**a**) and max duration (**b**) [14].

3. Results

With e-Genius-Mod a flexible test platform is realized which can be perfectly used for flight testing focused at academic research. The platform covers different fields of future research at the Institute of Aircraft Design at University of Stuttgart. Thanks to the design under similarity requirements it is an investigation platform for free flight models used as a tool in aircraft design.

Through its modular design new technologies can be tested with small, fast and low cost changes on the aircraft. The modular design allows changes while maintaining the baseline system design. In this way new research projects can focus on technology assessment in flight. With the platform a technology demonstration in relevant environment is partially possible.

Furthermore, the high payload capacity (Table 2) combined with a large free volume within the airframe allow an easy integration of system technologies and in flight experiments.

Table 2. Technical Data of e-Genius-Mod [11].

Wingspan	5.62 m
Aspect ratio	20.16
Length	2.95 m
MTOM	41 kg
Electric drive power	5 kw
Payload	up to 10 kg
Design speed	24.8 m/s
Max. speed	35.6 m/s
Airframe	2 Segments
Wing	8 Segments

3.1. Flexible Airframe Configuration

The e-Genius-Mod airframe parts are divided in segments (Figure 9), leading to multiple advantages. One aspect is the transportability of the aircraft system, which allows the integration and maintenance of the technology to be tested in the lab while being mobile enough to use different airfields for flight testing.

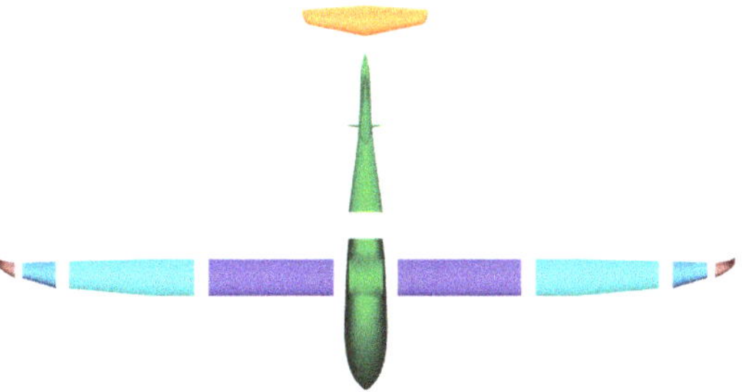

Figure 9. Segment overview e-Genius-Mod.

Furthermore, the segmentation enables the realization of drastic changes to the aircraft configuration. For example, new wings/airfoils can be tested without making changes to the fuselage segment. Another conceivable change in the configuration is the use of a modified tail segment that would allow for example an easy and fast assessment of a V-tail. In this case, only the tail segment would have to be replaced.

This was the motivation to divide this fuselage in two parts and an elevator. The wing consists of eight individual segments. Each wing half is assembled of two main segments which include the control surfaces, an outer surface to increase the aspect ratio and the winglet as wing tip. The segments have connection interfaces for data and power transmission.

3.2. Systems Design

The flight system of the e-Genius-Mod is designed for remote-controlled (RC) flight as well as automatic flight. The main part of the e-Genius-Mod is the flight management system, realized with a Pixhawk autopilot.

Complete radio coverage between pilot and aircraft, through all flight attitudes is guaranteed via two RC receivers with two antennas, at any time. Three antennas are installed around a fuselage section with an offset of 120° and one antenna is oriented lengthwise of the fuselage.

For fail safe reasons each segment of e-Genius-Mod (Figure 10) is connected with two buses of the flight management system to ensure redundancy. E-Genius-Mod has two separate data buses and two power buses. The power buses are realized with a Powerbox-System which is connected to two batteries for a redundant power supply. As a result, no single failure in the power supply system leads to a loss of the aircraft.

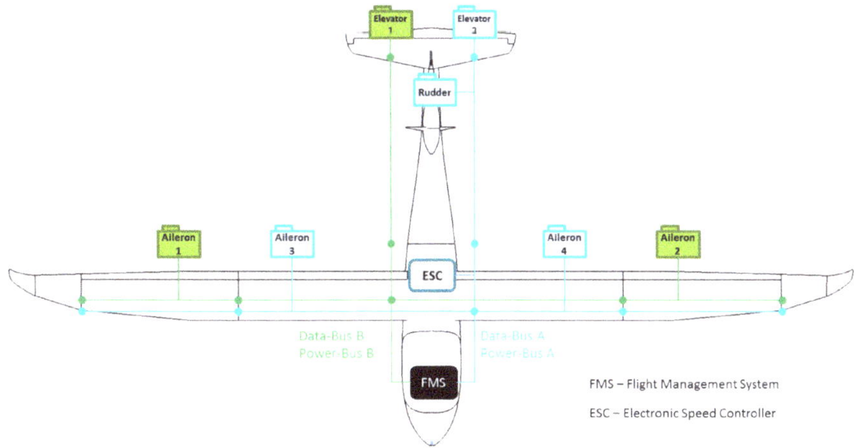

Figure 10. System overview e-Genius-Mod.

Even if one of the two data bus-systems failed, the actuators are distributed on both bus systems according to the following schematic, which still allows a sufficient control authority over the aircraft:

The full-scale e-Genius is designed with flaperons. These control surfaces are divided for the e-Genius-Mod. In doing so, smaller actuators can be used and the system reliability against failure of an activator is improved. Furthermore there is the advantage to use the outer surface as outboard aileron and the inner as inboard aileron and flap. The flaperon is only needed for investigation in cases where similarity conditions between scaled and full-scale e-Genius are necessary.

The shown connection has the advantage that the aircraft remains controllable even with only one bus system available. By losing bus A the outboard ailerons and one control surface of the elevator are still available. In case bus B is lost the inboard ailerons/flaps, one control surface of the elevator and the rudder are available. Because the effectiveness of the inboard ailerons is lower compared to the outboard ailerons, the inboard ailerons are connected on the same bus as the rudder. This way both failure cases guaranty sufficient controllability.

A failure of the electronic speed controller and the electrical motor is defined as noncritical because the e-Genius-Mod can fly as sailplane due to its high aspect ratio.

3.3. Free–Flight Model e-Genius Mod

Future research of the e-Genius-Mod is focusing on investigation of new aviation technologies, configurations, and similarity requirements. Therefore it is equipped as "free flight wind tunnel". The standard integrated measuring equipment allows a complete data collection for identifying the aircraft performance. Difficulties in free-flying measurements, however, are the uncontrollable environmental conditions, which can influence the results of the flight tests. In order to keep the disturbance as low as possible, the scaling factor of the e-Genius-Mod is sufficiently low.

The standard set of measurement data consists of the basic data required for the flight control and additional measured values to support the flight control system, as well as input data for the ground station. The main sensors are:

- Inertial measurement unit—data of flight attitude (angle of orientation, acceleration, rotation rate)
- Air data boom—angle of attack, side slip, static and dynamic pressure
- GPS—position of the aircraft
- Actuator feedback—control surface angle
- Monitoring sensors—battery current/voltage/temperature, engine RPM/temperature

Due to the internal volume of the e-Genius-Mod, additional systems for data acquisition can be implemented as required to generate the optimal data set, depending on the in-flight experiment.

3.4. Application Example—Investigation of a Wing Tip Propulsion System

As stated before, the flexibility of the UAS test bed and its modular design is well suited for investigation and demonstration platform for distributed propulsion systems. In a first step, a wing tip propeller will be investigated in the project ELFLEAN (Electric wing tip propulsion system for the development of energy-efficient and noise reduced airplanes), supported by Federal Ministry for Economic Affairs and Energy on the basis of a decision by the German Bundestag.

One aspect of the project is to test the wing tip propulsion system in flight on the UAS test bed e-Genius-Mod. For a more efficient investigation of the effects of wing tip propulsion, the aspect ratio of the UAS shall be reduced to increase induced drag.

Through the modular concept these requirements can easily be fulfilled. As shown in Figure 11, the outer segments of the wing can be disconnected to have a ready to fly reference configuration for the investigation.

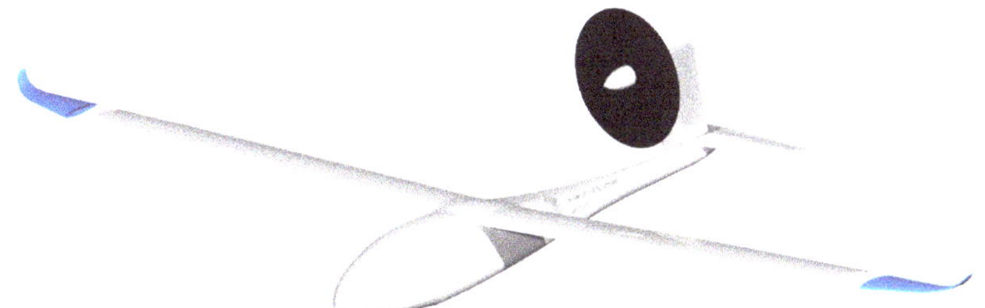

Figure 11. e-Genius-Mod with disconnecting wing tips to reduce aspect ratio.

Also, the connection of the wing tip propulsion system can be realized (Figure 12) with only one new lay-up mold for the pods of the system. An interface to connect the pods to the structure and the flight management system is already available on the shortened end of the wing.

Figure 12. e-Genius-Mod with integrated wing tip propulsion system.

4. Discussion

With the platform e-Genius-Mod a tool is available at the University of Stuttgart. For technology improvements at the full-scale, its scale is a perfect test platform. Because of its modular construction e-Genius-Mod it is an ideal tool to integrate and to test new technologies in-flight.

The UAS test platform e-Genius-Mod currently is used to test and validate various effects of distributed propulsion (Figure 13). Beyond this, it is planned to use the platform for all kinds of technologies, such as structural concepts, acoustical analysis, flight guidance and navigation or data transmission technologies. This will further demonstrate the value of modular and versatile UAS platforms for the early, safe and cost effective flight testing of innovative technologies and concepts.

Figure 13. Design study "Investigation of distributed propulsion".

5. Conclusions

The e-Genius-Mod as a free-flight demonstrator creates excellent opportunities for flight and system testing as well as the investigation of new aircraft configurations in terms of feasibility. Through its development in respect of Froude number e-Genius-Mod has the advantage of the comparability with its original model, system laws for flight mechanical and dynamical behaviors can be improved. Its modular design makes it possible to test different configurations and new aviation technologies (like new propulsion systems) time and cost efficiently. The unmanned aerial system e-Genius-Mod is a handsome tool to close the gap between upstream research and flight demonstration to reach a higher technical readiness level. For the Institute of Aircraft Design at University of Stuttgart, the system is a game changing advanced tool to test new aviation technologies and to expand the results of research and development tools like wind tunnel tests and CFD calculations.

Author Contributions: Conceptualization, D.P.B.; airfoil scaling, A.B.; variable pitch propeller, L.K.; writing—original draft preparation, D.P.B.; writing—review and editing, D.P.B., J.D., A.S.; supervision, A.S.; project administration, D.P.B.

Funding: This research received no external funding.

Conflicts of Interest: The authors declare no conflict of interest.

References

1. Bergmann, D.; Denzel, J.; Strohmayer, A. UAS as flexible and innovative test platform for aircraft configuration and systems testing. In Proceedings of the 8th EASN-CEAS International Workshop on Manufacturing for Growth & Innovation, Glasgow, UK, 4–7 September 2018.

2. Kittmann, K. Entwicklung einer modularen Messplattform zur Analyse des Potenzials von Freiflugmessungen. Ph.D. Thesis, University of Stuttgart, Institute of Aircraft Design, Stuttgart, Germany, 2013. [CrossRef]

3. Wolowicz, C.H.; Bowman, J.S., Jr.; Gilbert, W.P. *Similitude Requirements and Scaling Relationships as Applied to Model Testing*; NASA Technical Paper 1435; National Aeronautics and Space Administration, Scientific and Technical Information Branch: Washington, DC, USA, 1979.

4. European Commission; Directorate-General for Research and Innovation; Directorate General for Mobility and Transport. *Flightpath 2050 Europe's Vision for Aviation, Report of the High Level Group on Aviation Research*; Publications Office of the European Union, European Union: Luxembourg, 2011; ISBN 978-92-79-19724-6. [CrossRef]

5. IATA Technology Roadmap 2013, 4th Edition. 2013. Available online: https://www.iata.org/whatwedo/environment/Documents/technology-roadmap-2013.pdf (accessed on 30 November 2018).

6. Schmollgruber, P.; Gobert, J.-L.; Gall, P.-E.; Goraj, Z.; Jentink, H.W.; Näs, A.; Voit-Nitschmann, R. An innovative evaluation platform for new aircraft concepts. *Aeronautical J.* **2010**, *114*, 451–456.

7. Jordan, T.; Langford, W.; Hill, J. Airborne Subscale Transport Aircraft Research Testbed—Aircraft Model Development. In Proceedings of the AIAA Guidance, Navigation, and Control Conference and Exhibit, San Francisco, CA, USA, 15–18 August 2005. [CrossRef]

8. Jouannet, C.; Lundström, D.; Amadori, K.; Berry, P. Design of a Very Light Jet and a Dynamically Scaled Demonstrator. In Proceedings of the 46th AIAA Aerospace Sciences Meeting and Exhibit, Reno, NV, USA, 7–8 January 2008. [CrossRef]

9. Chambers, J.R. *Modeling Flight, The Role of Scaled Free-Flight Models in Support of NASA's Aerospace Programs*; NASA Technical Report, National Aeronautics and Space Administration; NASA Headquarters: Washington, DC, UAS, 2010; ISBN 978-0-16-084633-5.

10. Nguewo, D. Erstellung und Optimierung der Skalierungsgesetze zur Abschätzung der Aerodynamik und der Eigendynamik eines Flugzeugs auf der Basis von frei fliegenden Modellen. Ph.D. Thesis, University of Stuttgart, Institute of Aircraft Design, Stuttgart, Germany, 2007. [CrossRef]

11. Baden, A. Entwurf eines Freiflugmodells des E-Motorseglers e-Genius unter Berücksichtigung der aerodynamischen Vergleichbarkeit. Bachelor's Thesis, University of Stuttgart, Institute of Aircraft Design, Stuttgart, Germany, 2016.

12. van Gorcum, J.J. Ready for Take-off: Scaled Flight Testing—An investigation into influence of scaling on the aerodynamic properties. Master's Thesis, Delft University of Technology, Delft, The Netherlands, 2017.

13. Lutz, T. *Profilentwurf, Skript zur Vorlesung und Seminar*; Lecture Notes; University of Stuttgart, Institute of Aerodynamics and Gas Dynamics: Stuttgart, Germany, 2007.

14. Kugler, L. Analysis and characterization of the variable pitch propeller for the Scaled-down e-Genius. Bachelor's Thesis, University of Stuttgart, Institute of Aircraft Design, Stuttgart, Germany, 2018.

15. McCrink, M.H.; Gregory, J.W. Blade Element Momentum Modeling of Low-Re Small UAS Electric Propulsion Systems. In Proceedings of the 33rd AIAA Applied Aerodynamics Conference, Dallas, TX, USA, 22–26 June 2015. [CrossRef]

 *aerospace*

Article

Verification of Boundary Conditions Applied to Active Flow Circulation Control

Petr Vrchota , **Aleš Prachař *** , **and Pavel Hospodář**

VZLU—Czech Aerospace Research Centre, 199 05 Prague, Czech Republic; vrchota@vzlu.cz (P.V.); hospodar@vzlu.cz (P.H.)
* Correspondence: prachar@vzlu.cz; Tel.: +420-225-115-192

Received: 30 November 2018; Accepted: 4 March 2019; Published: 8 March 2019

Abstract: Inclusion of Active Flow Control (AFC) into Computational Fluid Dynamics (CFD) simulations is usually highly time-consuming and requires extensive computational resources and effort. In principle, the flow inside of the fluidic AFC actuators should be incorporated into the problem under consideration. However, for many applications, the internal actuator flow is not crucial, and only its effect on the outer flow needs to be resolved. In this study, the unsteady periodic flow inside the Suction and Oscillatory Blowing (SaOB) actuator is analyzed, using two CFD methods of ranging complexity (URANS and hybrid RANS-LES). The results are used for the definition and development of the simplified surface boundary condition for simulating the SaOB flow at the actuator's exit. The developed boundary condition is verified and validated, in the case of a low-speed airfoil with suction applied on the upper (suction) side of the airfoil and oscillatory blowing applied on the lower (pressure) side, close to the trailing edge—a fluidic Gurney flap. Its effect on the circulation is analyzed and compared to the experimental data.

Keywords: computational fluid dynamics; active flow control; suction and oscillatory blowing actuator; circulation control

1. Introduction

Active Flow Control (AFC) is considered to be a useful technique in enhancing the performance of existing designs. However, it is challenging to include AFC when already in the design phase. The reason for this is the high geometric complexity of the AFC flow actuators (jets, suction holes, internal structure, and tubing) which should be incorporated, in the case of full resolution. Consequently, the required resources increase due to various reasons. First, it adds complexity to the computational grids, which, in the case of full resolution, need to include the internal shape of the actuators themselves. Furthermore, a heavy workload is needed during the grid generation process, which is far from being automatic. A second, and in some sense complementary, issue is the growth of the problem size in terms of the number of grid points, which leads to increased wall-time and Central Processing Unit (CPU) resources needed to solve the problem. This is even more significant in the case of actuator arrays, where even the effort to simulate the AFC internal flows easily exceeds the typical size of the corresponding problem without AFC. Moreover, the flow is usually inherently unsteady by nature, especially in the case of pulsed or oscillatory blowing, which requires time-consuming and unsteady calculations, as well as a statistical evaluation of the results. Moreover, there might be a large time-scale separation between mean flow phenomena and the AFC actuator internal flow frequencies. While it is desirable to simplify AFC in the Computational Fluid Dynamics (CFD) simulation (i.e., to simulate the effect of AFC, rather than to spend resources on the physical modeling of the AFC system), this cannot be done without sufficient knowledge about the internal flow of the actuator or the character of the actuator exit flow [1,2].

It has become quite popular to control flows in many aeronautical, transport, and industrial applications [3–5]. A number of computational studies rely on a uniform flow field at the AFC device exit [6–10]. It is evident that a realistic jet exit flow field is needed, in order to define appropriate boundary conditions [2,11,12] and to obtain physically consistent interactions between the actuator and the outer flow. This can be done by modeling the actuator geometry, or at least a significant part like the actuator exit nozzle, to enable development of the velocity profile and to obtain a more realistic flow field at the exit from the actuators [13–16]. The functional representation for the nozzle oscillatory velocity profiles of the SaOB (Suction and Oscillatory Blowing) actuator was developed in [12], which can be used as surface boundary conditions for complex AFC simulations. Although this representation was only 2D, it matched the measured velocity profile at the nozzle's exit quite well. Moreover, a proper orthogonal decomposition, to create a reduced-order model of the actuator's effect on the flow-field, was also studied [17] with the aid of Particle Image Velocimetry (PIV) experimental data.

Our goal is divided into several consecutive steps. First, we study the unsteady internal flow in a single SaOB actuator [18]. This is done by the CFD simulation using an Unsteady Reynolds Average Navier–Stokes (URANS) solver with a k–ω type turbulence model [19], and using hybrid Reynolds Average Navier–Stokes (RANS)-Large Eddy Simulation (LES) turbulence modeling [20]. The use of two approaches with different levels of complexity, validated using the experimental data [21], allows better insight and confidence in the computational methods. This approach, although still quite heavy on resources, is an alternative to the even more demanding LES simulations [22], which require a computational force that is out of reach, in most cases.

Based on the data from these CFD simulations, the unsteady inflow boundary condition for the simulation of the SaOB actuator has been defined, as briefly described previously in [23]. It is based on the use of Gaussian functions to obtain a spatial representation of the velocity field, combined with a harmonic function to represent the time evolution. The boundary condition, instead of the complex internal structure of the actuators, further enables conservation of the complexity of the problem under consideration, in a reasonable range. It is possible to save the load of millions of grid points by substitution of the internal structure of the actuator by the boundary condition at the actuator's exit. It is also possible to consider actuator arrays and the incorporation of AFC during the design process, by this approach.

The second level of validation is presented by considering a subsonic flow around a quasi-2D airfoil with an integrated SaOB actuator. The increase of the lift was observed with the AFC active. The comparison of our CFD results with the experimental data [21] shows that the effect of the SaOB actuator was replicated in the simulation.

The structure of the paper is as follows. In Section 2, the principle of the SaOB actuator is briefly described, the computational problem is defined, and the CFD methods are presented. The data extracted from the CFD simulations are used, in Section 3, for the definition of the surface boundary condition. The mathematical background is briefly explained and the level of agreement between the data and the analytical description is presented. Section 4 covers the application of the simplified boundary condition to the case of a quasi-2D airfoil to enhance its circulation. The calculated values are compared to the experimental data. The conclusion briefly sums up the achieved results, and outlines other possible applications.

2. Characterization of the SaOB Actuator

Oscillatory blowing is an effective tool for increasing the momentum of the flow inside the boundary layer—which can be used, for example, for flow separation control. Periodic blowing through a series of narrow slots enhances the shear-layer mixing and transfer of high-momentum fluid from the outside of the shear layer to the near wall region, which delays the boundary-layer separation. In the present paper, the actuator is used as a fluidic Gurney flap to enhance airfoil circulation.

The principle of the SaOB actuator, as in [18], is described in Figure 1.

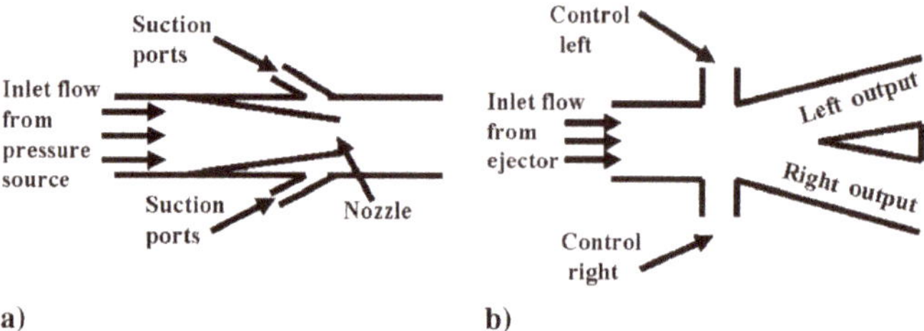

Figure 1. Schematic rendering of the Suction and Oscillatory Blowing (SaOB) actuator: (**a**) Ejector, and (**b**) switching valve (adopted from [18]).

The actuator is a combination of an ejector and a bi-stable fluidic amplifier. When a jet stream from the pressure supply is ejected into a bigger conduit, a low-pressure region is created, due to entrainment (Figure 1a). The cavity behind the jet is open to the outer environment, through the suction ports. As a result, the pressure gradient between the internal jet area and the external environment causes air to be sucked into the cavity. This additional mass flow depends on the actual pressure gradient. This, on one hand, increases the total mass flow through the actuator and, on the other hand, it naturally causes suction—which can be further exploited by positioning the suction holes in a desirable area. Moreover, only the pressure source is needed. The oscillations of the flow between the right and left output slots (Figure 1b) is caused by the natural flow switching, controlled by the flow through the feedback tube connecting the control ports (control left and control right, see Figure 1b). No moving parts are needed to obtain oscillation.

The SaOB actuator is very effective, especially in the low range of the blowing momentum coefficient, C_μ (see Equation (3)), in comparison with other types of fluidic actuators (steady, pulsed blowing, sweeping jets, and so on). It can be used for many applications, from flow separation control to maneuvering [21,24].

2.1. CFD Simulation of the Isolated SaOB Actuator

The CFD simulations were focused on the internal flow-field and on the flow at the nozzle exit of the single SaOB actuator. These simulations corresponded to the bench-top test of the SaOB in the laboratory, in still-air ambient conditions [21]. The geometrical setup of the detailed actuator interior layout is shown in Figure 2.

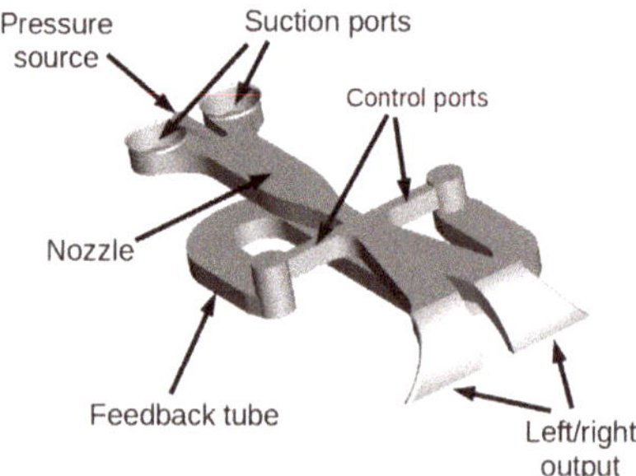

Figure 2. The geometry of the SaOB actuator; the shade color represents the adiabatic wall boundary conditions.

As we can see, a realistic geometry of the SaOB actuator was utilized, and its geometrical details were fully resolved. In terms of boundary conditions, the pressure supply and suction ports were represented by the inflow boundary conditions. The total pressure, total temperature, and flow direction were prescribed at these boundaries, a boundary condition usually denoted as total states inlet or stagnation pressure condition. Flow direction was set as perpendicular to the geometry of the boundary conditions. The total pressure at the pressure supply boundary was varied, to obtain a wide range of working points for the comparison between CFD computations and the experimental data (available from [21]). The ambient atmospheric conditions were prescribed at the suction ports. The exit nozzles of the SaOB actuator were connected to the rectangular domain, which simulated the flat plate of the wind tunnel wall. To consider the effect of neighboring actuators, periodic boundary conditions were prescribed at the sides of the computational domain. The SaOB walls were represented by the adiabatic no-slip wall conditions. The outer flow freestream velocity was set to a value close to zero, to mimic the no-flow condition of the experiment, and then to 25 m/s, to study interaction with the outer flow. The boundary conditions in the CFD solver were implemented as weak boundary conditions; that is, they were imposed by the boundary flux, rather than assigned directly.

The internal grid of the SaOB actuator and of the computational domain was created in the Pointwise grid generation software. The grid was refined in the regions of interest (ejector, output nozzles, and the region in the vicinity of the actuator outputs). The hybrid grid grew from the mostly quad-dominant surface grid into the volume, to create a sufficiently fine grid representation in the boundary layer, and the unstructured tetra cells filled the rest of the volume. Due to unsteady nature of the flow and the large variation in the local velocities, it was challenging to set grid parameters a priori. Therefore, the grids were tuned according to preliminary results, to provide acceptable convergence, and $y+ < 1$, as recommended for the use of turbulence models. The computations were very expensive, in terms of computational time and resources. Therefore, a systematic grid convergence study is not presented. However, since the problem was solved in two steps (URANS and SST-DDES on a heavily-refined grid), the consistency of results is considered as a partial substitute.

The number of grid points is presented in Table 1, where the principal parts of the actuator grid are highlighted. The first cell height range was due to different settings in the centre of the SaOB central channel and in the control ports, where the velocity was much lower. The growth rate from the first cell height was set close to 1.2. While the grid had about 5.5 million grid points for the URANS campaign [25], it was uniformly refined—starting from the finer surface grid—to about 12 million, just inside the actuator, for the SST-DDES turbulence model study [26]; see Figure 3.

Table 1. Grid parameters for the URANS and SST-DDES grid cases.

Grid Part	URANS Grid		SST-DDES Grid	
	# Points	First Cell (mm)	# Points	First Cell (mm)
Total	5,418,072	–	20,770,098	–
Feedback tube	398,033	0.01	3,882,487	0.003
Nozzle and outputs	2,668,014	0.01–0.001	7,819,016	0.008–0.001

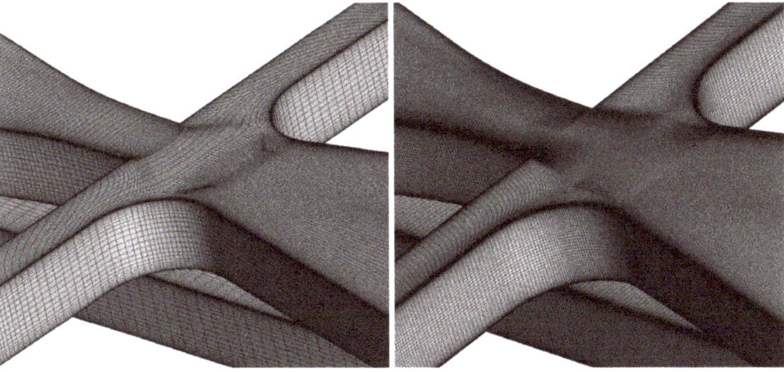

Figure 3. Detail of the surface computational grids for the URANS calculations with 5.5 million grid points (**left**) and for the hybrid SST-DDES (**right**), refined to nearly 21 million grid points in total, with 12 million inside the actuator. See Table 1 for details.

All simulations were carried out in a Finite Volume Navier–Stokes solver, Edge [27]. It was designed as a solver for the Navier–Stokes equations on unstructured meshes. The data structure inside the solver is edge-based, and the code was constructed as cell-vertex. It employs multi-grid and residual smoothing acceleration for steady-state problems. Moreover, the time accurate (unsteady) simulations use an implicit dual-time stepping approach, where, in each time step, a steady-like problem is converged to with a sequence of inner iterations. The maximum number of inner iterations was determined by the drop in the density residual (2 orders of magnitude), or by the maximum number of inner iterations reached (100–300, depending on the case). The spatial discretization of the convective terms relied on the Jameson-Schmidt-Turkel (JST) scalar scheme [28], which is of the second order of accuracy.

All simulations were run as a fully turbulent (i.e., no laminar region or laminar-to-turbulent transition was prescribed). The turbulence model used for the URANS simulations was Wallin & Johansson's EARSM + Hellsten k–ω model [19], and the SST-DDES model, as presented in [20], was employed for the hybrid RANS-LES calculations.

The physical time step was varied for particular cases, depending on the oscillation frequency, to reach an adequate number of time steps per cycle of the blowing velocity, from one actuator's outlet the other, while switching.

This was defined by the sampling rate. The sampling rate of the oscillation frequency was kept at a similar level for all simulated cases. Its value was high enough to obtain a smooth description of the oscillating frequency. This means that the physical time step was varied from 0.00002 s, corresponding to the lowest oscillation frequency, to 0.00001 s, for the highest oscillation frequency. This gave approximately 200 time steps per oscillation period. The typical snapshot of the flow inside the actuator is presented in Figure 4.

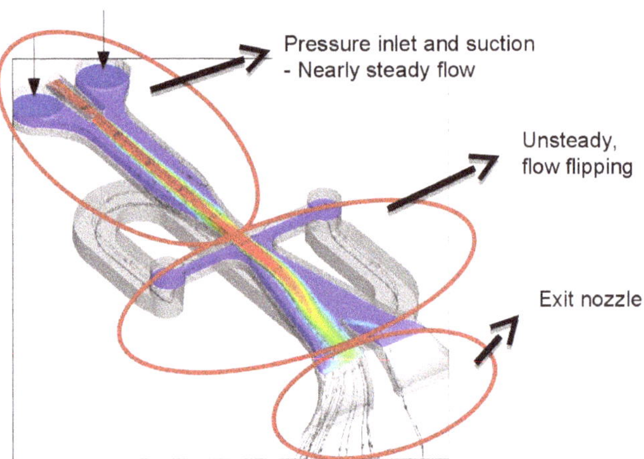

Figure 4. Snapshot of typical flow inside the SaOB actuator. The main areas, with respect to the character of the flow, are highlighted, adopted from [26].

2.2. CFD Results and Interpretation

The main aim of this study was to obtain the time-dependent flow field variables at the SaOB actuator exit nozzle for different supply pressures. These results serve as a source of data for boundary condition definition. Furthermore, comparison between URANS with the k–ω model and the hybrid RANS-LES, model represented by the SST-DDES simulation, brings more confidence in the computational results.

A basic comparison can be made using the experimental results (as in [21]), where the actuator exit peak velocity and oscillation frequency were evaluated; see Figure 5. The comparison of the maximum outlet nozzle velocity and the oscillation frequency, which naturally develops due to the flow inside the actuator, is presented. The frequency range of between $f = 170$ Hz and $f = 440$ Hz was obtained; which relates to the maximum actuator outlet velocities, in the range between $V_{max} = 24$ m/s and $V_{max} = 56$ m/s.

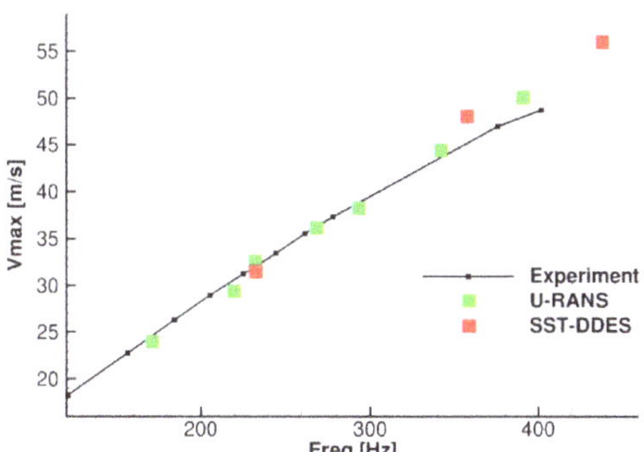

Figure 5. Comparison of the Computational Fluid Dynamics (CFD) results with the experimental data [21].

A good agreement between the experimental data and the computational results of the URANS and hybrid RANS-LES simulations was achieved. It has to be noted that there was a certain degree of non-symmetry in the results (between the left–right actuator outputs) that was observed both in the experimental data [21] (where it can be attributed to the manufacturing tolerances of the actuators) and also in the CFD results, which is probably due to a not fully-symmetric grid and the S-shaped feedback channel. The data were averaged from both output nozzles before they were presented. More details on the evaluation of the data are given in Section 3. The linear trend between the oscillation frequency and the peak velocity at the nozzle exit was preserved for all the results mentioned here.

The velocity profile at the actuator's exit was highly non-uniform in both directions (streamwise and spanwise), see Figure 6. This was not necessarily a drawback, since it enhanced mixing inside the boundary layer. As the velocity profile was subject to a complex flow field inside the actuator, it was necessary to accurately simulate the internal flow, capture local separations, and so on. In our case, the non-uniformity of the flow could also be attributed to the shape of the exit nozzle (see Figure 4), which deflected the flow from the streamwise direction to the perpendicular normal direction. The flow was, thus, due to its momentum attached to the outer wall, along the outlet nozzle radius.

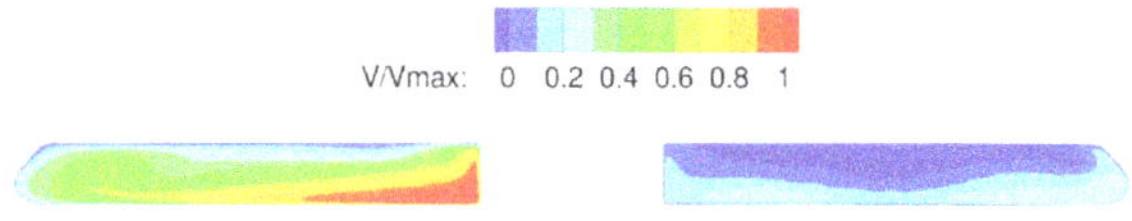

Figure 6. Normalized velocity at the actuator exit at the peak velocity phase (left nozzle), adopted from [26].

Figure 7 shows a comparison of the SaOB actuator exit nozzles at the time instant when the peak value V_{max} was achieved. While the URANS results were spatially interpolated (due to probe recording done during the simulations), the hybrid RANS-LES results showed instantaneous values from the cut through the computational domain in the same spatial position. There was no significant deviation between both peak velocity profiles, although it is possible to see some differences in the velocity profile of the 'inactive' actuator outlet (left outlet, in the case that the right outlet is in the peak velocity phase, and vice versa). The residuals of the flow through the nominally inactive outlet nozzle contributed to the nonzero mass flux through this outlet. However, the momentum there was relatively small. Another nice property of the velocity profile is that it was fairly self-similar for different operational states of the SaOB actuator; see Section 3. This has been observed not only in [23], for the present version of the SaOB actuator, but also in [22], for a slightly different geometry.

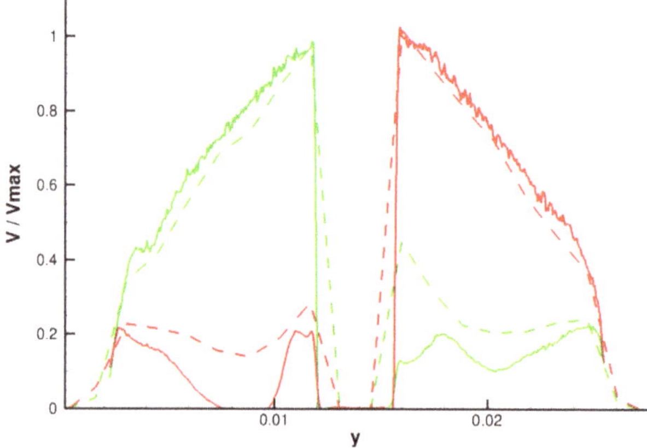

Figure 7. Comparison of exit velocity profile, normalized by the maximum velocity. The URANS results (dashed) and SST-DDES simulations (solid) are shown. Green lines represent the peak velocity snapshot for the left outlet nozzle, and red lines represent peak velocity for the right nozzle.

3. Definition of the Boundary Condition

The time-dependent flow-field variables at the actuator's exit were extracted from the CFD simulations and approximated by the surrogate model. The main character of the flow was periodic with the frequency of the flow oscillation, corresponding to different inlet pressures into the SaOB actuator.

A wide range of usage of surrogate modeling had been seen in the field of computational science in recent decades. There are many mathematical approaches that provide surrogate modeling: Kriging, Response surface method [6], artificial neural networks [29], Gaussian processes [30], and so on. In this case, an analytical function, describing the character of the flow field, was used. More specifically, the normal and spanwise components of velocity were described as a sum of the two-dimensional Gaussian functions multiplied by the sine function; the generic form is:

$$V(x, y, t) = \sum_{i=1}^{N} A_i e^{-\left(\frac{(x-x_i)^2}{2\sigma_{x,i}} + \frac{(y-y_i)^2}{2\sigma_{y,i}} \right)} \cdot \left(sin(2\pi f t + \phi_i) + c_i \right). \tag{1}$$

The Gaussian itself is defined by the five parameters $(A_i, x_i, y_i, \sigma_{x,i}, \sigma_{y,i})$, and is related to the spatial distribution of the velocity profile. The character of the oscillations is described by the sine function, specified by an additional three parameters with obvious physical meanings (oscillation frequency f, phase shift ϕ_i, and mean value c_i). The comparison between the calculated and approximate velocity at the nozzle exit is depicted in Figure 8. The numbered signs depict the locations of the Gaussians. The difference between the CFD and the approximated data was very small and could be neglected at this stage.

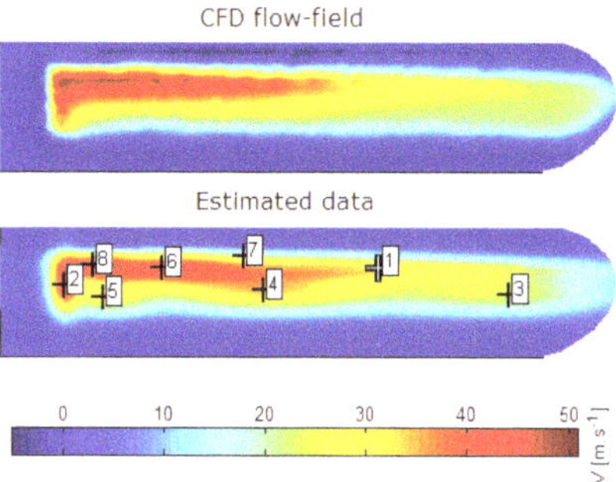

Figure 8. Comparison of the calculated and approximated vertical velocity components, corresponding to the same time instant. A snapshot of the CFD flowfield (**top**) is compared to the approximation using Gaussians (**bottom**). The numbered marks represent the coordinates $(x_i, y_i), i = 1, \ldots, 8$, as in Equation (1).

The velocity distribution from the CFD simulations, for various peak outlet velocities of the SaOB actuator, are shown in Figure 9. It shows how the values of maximum outlet velocities were extracted for the evaluation, as in Figure 5. It can also be seen that the shape of the velocity profile was essentially independent of the particular case (i.e., inlet pressure to the SaOB actuator), and that the velocity profiles are reasonably self-similar.

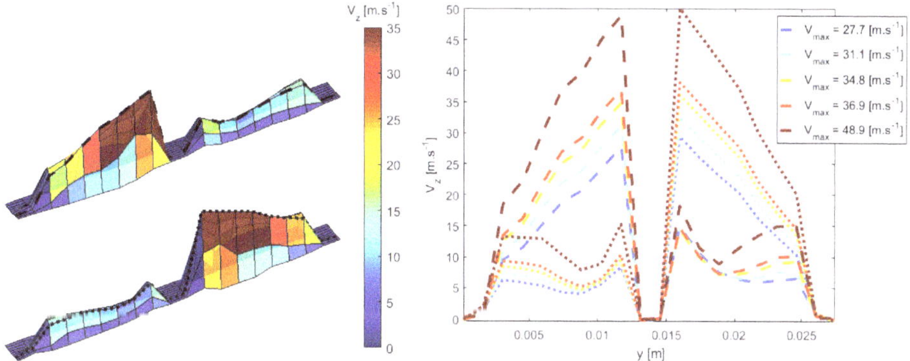

Figure 9. Determination of the peak velocity from the surface interpolation of the velocity field at the nozzle exits. The left figure displays velocity profiles at the peak velocity time instant in the right and the left outlet nozzles of the SaOB actuator, respectively. The dashed and dotted black lines show a spanwise cut through the maximum velocity location. The graph on the right displays velocity profiles at the peak velocities of (approximately) 28 m/s, 31 m/s, 35 m/s, 37 m/s, and 49 m/s (colors are according to line legend). The dashed lines represent velocity peaks at the left nozzle, dotted lines are the right nozzle peaks.

An optimization procedure was defined to obtain parameters that provided a minimum value of the cost function. Two parts of the cost function were defined to fulfill the desired characteristics of time dependent flow-field (see Figure 10) and total volume flow rate through the actuator. Therefore, the cost function is defined as:

$$\min_{\theta} J = a \sum_{j=1}^{k} \left(V_j^*(\theta) - V_j \right) + b \sum_{j=1}^{k} \left(Q_j^*(\theta) - Q_j \right), \qquad (2)$$

where $V_j^*(\theta)$ and V_j are estimated and computed flow-field variables (velocities), respectively; $Q_j^*(\theta)$ and Q_j are estimated and computed volume flow rates, respectively; k is the number of samples (time instances); a and b are optimization weights; and θ is a design vector describing the flow variable, as in Equation (1). In this case, only eight steps ($N = 8$) were used, which we considered sufficient, given the harmonic nature of the oscillations. The coefficient of determination $R^2 = 99.15\%$ for the mass flow evolution fit between the CFD data and the analytical function, which suggests that the process was described with sufficient accuracy.

Three different oscillation frequencies, across the entire range tested during the bench-top test, were calculated by this boundary condition for verification purposes. The courses of the volume flow rate, based on the vertical component of the blowing velocity, dependent on time, are depicted in Figure 10. The solid lines show the volume flow computed by the CFD, and the dashed lines are based on the prescribed values of the boundary condition. It was found that this functional representation enables the use of several parameters, common for the particular SaOB actuator, and only a few (the frequency f, and the amplitudes A_i) needed to be adjusted. Moreover, a linear dependence between the frequency and amplitudes was observed for this range of operational states (i.e., when the frequency of the oscillations is altered, only the amplitudes A_i needed to be uniformly scaled). This scaling had to be determined, based on the evaluation of a range of cases.

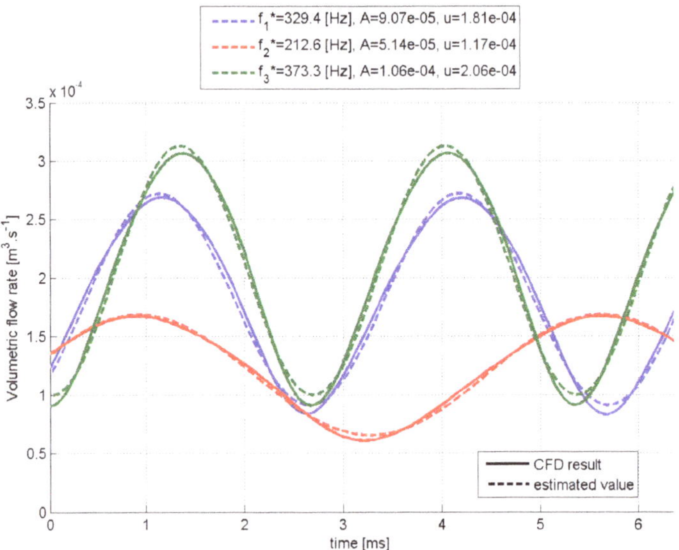

Figure 10. Volume flow rates—calculated and prescribed by the boundary condition model—for three different SaOB working states.

4. Application to Circulation Control

The boundary condition with variable velocity profiles, as defined in Section 3, was applied to the low-speed airfoil, AR2, with the wind tunnel conditions according to [21]. Both possible regimes of the SaOB actuator were evaluated, namely full (including both suction and oscillatory blowing parts) and without suction (only oscillatory blowing). The experimental data were used for comparison with the CFD results presented here.

The computational domain consisted of one spanwise segment of the airfoil, incorporating only one pair of the SaOB outlet nozzles. The domain was set as periodic in the spanwise direction to simulate a row synchronized actuators, and it was extended to approximately 100 airfoil chords in the streamwise and normal directions. The suction was placed at 82% of the chord (on the upper surface of the airfoil) and the location of the blowing slots was at 95% of the chord (on the lower surface, near the trailing edge), see Figure 11.

Figure 11. Detail of the CFD surface grid around trailing edge of the AR2 airfoil. The cyan surface represents blowing slots, the green color marks the circular suction holes.

The computational grid had 5,978,534 nodes in total. The airfoil was resolved with approximately 800 points in the streamwise direction (around whole airfoil) with a fine resolution close to the leading edge, trailing edge, and near the positions of the suction and blowing slots. It was tuned to achieve the usual requirements for $y+ < 1$, with a Reynolds number of 600,000, based on the chord length and the freestream velocity $U_\infty = 25$ m/s. Figure 12 shows the spanwise cut through the grid in the area of the trailing edge in detail.

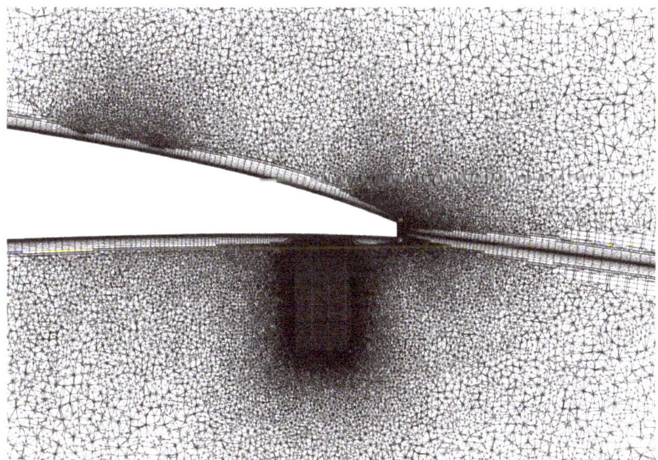

Figure 12. Detail of the spanwise cut through the CFD grid close to the trailing edge of the AR2 airfoil. A structured block, to resolve oscillatory blowing and the wake, is included.

4.1. CFD Simulations with the SaOB Boundary Condition

For the simulation of the effect of the oscillatory blowing part of the SaOB actuator, the velocity profile (defined by Equation (1)) was applied at the blowing slots, see Figure 11 (cyan). For the effect of the full SaOB, the additional surface boundaries for suction were added into the computational grid on the upper surface of the airfoil, see Figure 11 (green). The boundary conditions used at these suction holes prescribed mass flow, according to [31]. The parameters defined by Equation (1) were varied, to obtain the blowing velocity and oscillation frequency over the whole range of considered supply pressures used during the verification process.

The momentum coefficient was calculated, according to the equation:

$$C_\mu = \frac{S_{slot}}{S_{ref}} \left(\frac{U_j}{U_\infty} \right)^2 ,$$

(3)

where the area of the SaOB outlet slot S_{slot} and the maximum outlet velocity U_j are put into relation with the reference area of the airfoil S_{ref} and the freestream velocity U_∞. The cut in the plane, normal to the streamwise direction, is shown in Figure 13. The case of full SaOB at $f = 300$ Hz and of only the oscillatory blowing (i.e., with the suction ports inactive) are depicted. Small differences between the C_p values are visible, namely a slightly higher suction peak on the upper surface of the airfoil (the color map is reversed) is visible, and the stagnation point moved.

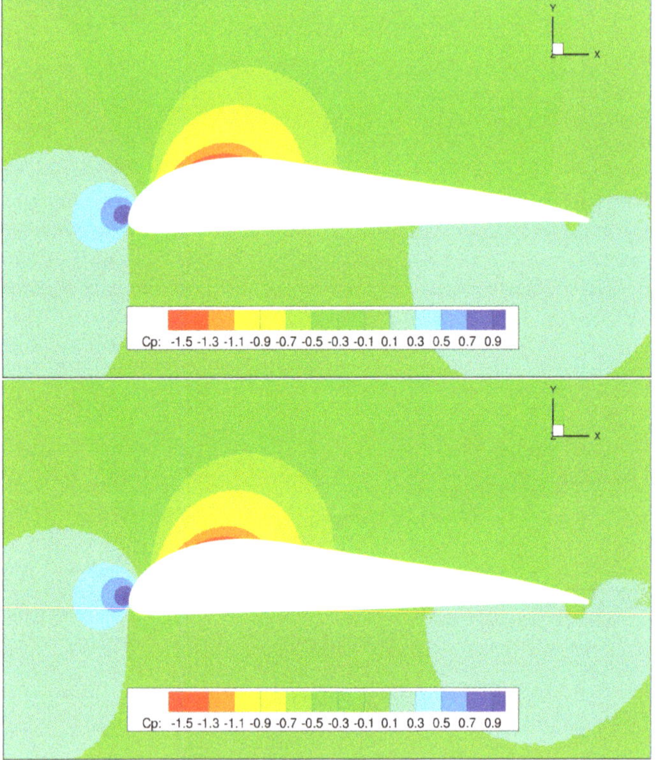

Figure 13. Comparison of full suction and oscillatory blowing (**top**) and plain oscillatory blowing (**bottom**) cases. Both simulations were run with an oscillation frequency of 300 Hz, and Angle of Attack (AoA) $= 0°$.

4.2. CFD Results and Experimental Data Comparison

The effect of ACF on ΔC_L, relative to the baseline configuration without any AFC, was evaluated and compared with the experimental data for Angle of Attack (AoA) = 0° [21,24], see Figure 14.

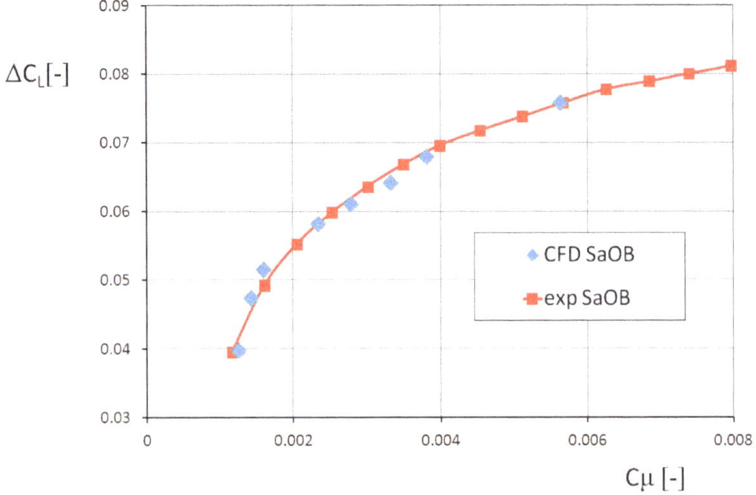

Figure 14. Comparison of the experimental data (taken from [21]) with the CFD results. Steady suction and oscillatory blowing actuator active.

The values of ΔC_L from the CFD simulations were in a good agreement with the experimental data. Although some differences still persisted, especially in the lower range of C_μ. These differences could be explained by small suction rates, where a small change could cause a relatively high effect.

The case of oscillatory blowing without added suction was simulated simply by switching the outflow boundary condition off and assigning the adiabatic wall boundary condition to the area of the suction ports. Figure 15 shows the integral values of ΔC_L, in relation to the C_μ.

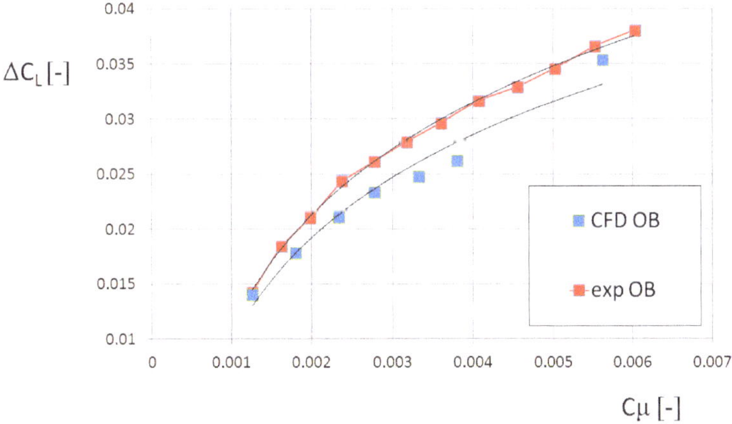

Figure 15. Comparison of the experimental data (see [21]), and the CFD results. Oscillatory blowing without suction. Black lines represent trend.

It is possible to see that ΔC_L was slightly underestimated between the minimum and maximum values of the considered C_μ range; however, the trend was very similar. It has to be mentioned that the differences between measured and calculated values were within $\Delta C_L < 0.005$. This difference could be the result of the different experimental and CFD calibrations. In CFD, the C_μ was given by setting the oscillatory blowing boundary condition, while, in the experiment, C_μ was correlated with supply pressure. Suction through the openings was considered during the bench-top test calibration. The suction flow increased the mass flow rate through the actuator's outlets, which resulted in the slightly higher blowing velocity—and, hence, a higher C_μ.

The local Cp distribution provides another way to assess the results. As was already seen in Figure 13, small differences can be expected. Figure 16 shows the systematic change of C_p due to suction (on the upper side of the airfoil), and also upstream of the oscillatory blowing (on the lower side of the airfoil). The oscillation frequency was used to denote specific cases; however, the maximum blowing velocity changes accordingly, as well. The detail of the trailing edge region Cp distribution is shown in Figure 17, where both the SaOB and oscillatory blowing are visible. It can be seen that the effect of oscillatory blowing (without suction) was limited mostly to the Cp variation on the lower side of the airfoil, upstream of the oscillatory blowing slots. This explains why the effect of SaOB was significantly higher, see Figures 14 and 15. This also shows how important it is to carefully select the positioning of the AFC devices.

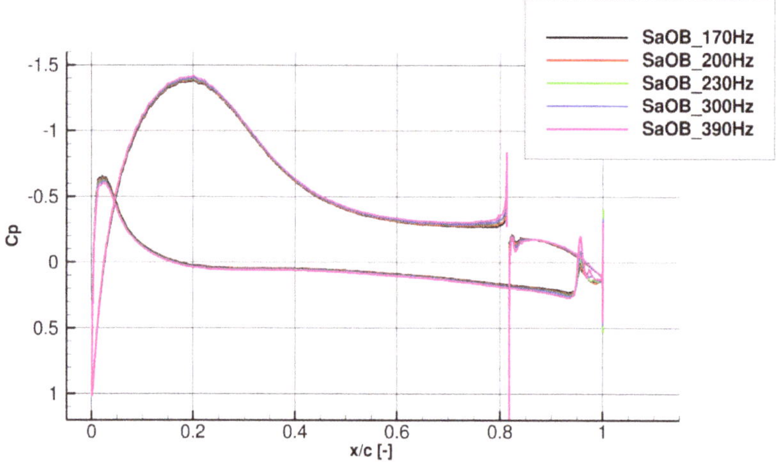

Figure 16. Cp distribution of the SaOB cases.

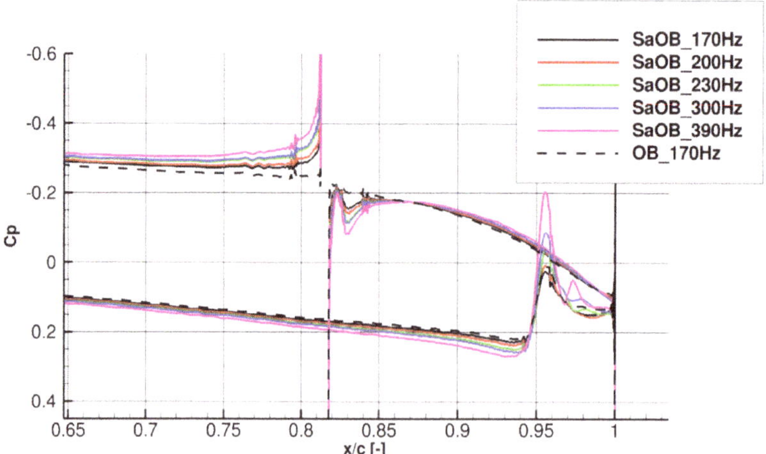

Figure 17. *Cont.*

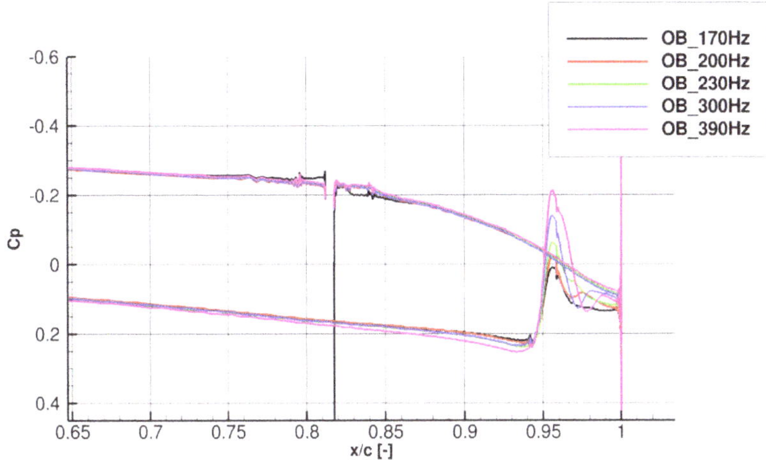

Figure 17. Detail of C_p distribution at the trailing edge region. SaOB case (**top**) and oscillatory blowing (**bottom**). Oscillatory blowing for low C_μ is added to the upper figure, for reference.

5. Conclusions

The time-dependent simplified boundary condition of the SaOB, based on the URANS simulations of the actuator's internal flowfield in still-air ambient conditions, has been defined. The developed boundary condition has been implemented into the CFD code, and verified for different blowing velocities and oscillation frequencies in the case of a low-speed airfoil, where steady suction and oscillatory blowing were applied at the upper and lower surfaces, respectively. Both cases have been simulated and validated against the experimental data. The obtained results are in a very good agreement with the experimental data; nevertheless, some differences in ΔC_L still exist, especially in the case of pure oscillatory blowing without suction. These differences in ΔC_L could be related to the calibration of both (experimental and CFD) methods. While the experimental data relate the blowing velocity to the supply pressure into the SaOB actuator, calibrated in still-air conditions, the CFD directly prescribes velocity components at the actuator boundary condition.

Although the simulations still need to be run in unsteady mode with a sufficiently small time step, applying this simplified boundary condition will allow for a reduction in grid points, by omitting the complex internal geometry of the actuator and, hence, save computational resources and time.

This simplified boundary condition will also enable efficient simulations of actuator arrays, used in low-speed as well as high-speed engineering applications, and will make the direct application of the AFC to the design process more feasible. Our follow-up work will be focused on the use of the presented boundary condition on additional typical aerodynamic applications, as in the case of separation control of the high-lift configuration with an ultra-high bypass ratio engine—the primary goal of the project under which the work has been performed.

Author Contributions: P.V. defined the problem, performed the calculations, extracted and analyzed the data, and provided parts of the manuscript. P.H. established the functional representation and provided parameters for the boundary condition definition. A.P. implemented the code, performed part of the calculations, analyzed the data, and finalized the manuscript.

Funding: The work presented in this paper and the research leading to these results has received funding from the European Commission under the Clean Sky 2 programme H2020-EU.3.4.5.1.—IADP Large Passenger Aircraft, grant agreement ID 754307, INAFLOWT project.

Acknowledgments: This work also was supported by The Ministry of Education, Youth, and Sports of the Czech Republic from the Large Infrastructures for Research, Experimental Development, and Innovations project IT4Innovations National Supercomputing Center LM2015070. Furthermore, the authors wish to thank our colleague Miroslav Šmíd for providing computational grids, and to Avraham Seifert and Danny Dolgopyat, who provided the experimental data.

Conflicts of Interest: The authors declare no conflict of interest.

Abbreviations

The following abbreviations are used in this manuscript:

AFC	Active Flow Control
AoA	Angle of Attack
CFD	Computational Fluid Dynamics
CPU	Central Processing Unit
C_μ	Jet momentum coefficient, $C_\mu = \frac{S_{slot}}{S_{ref}} \left(\frac{U_j}{U_\infty} \right)^2$
DDES	Delayed Detached Eddy Simulation
f	frequency (Hz)
LES	Large Eddy Simulation
PIV	Particle Image Velocimetry
RANS	Reynolds Average Navier–Stokes
SST	Menter's Shear Stress Transport (turbulence model)
S_{ref}	Reference area
S_{slot}	Outlet slot area
t	Time (s)
U_j	Maximum jet velocity (m/s)
U_∞	Free stream velocity (m/s)
URANS	Unsteady Reynolds Average Navier–Stokes
$V_{\max}$	Maximum velocity from the SaoB actuator nozzle (m/s)

References

1. Blaylock, M.; Chow, R.; Cooperman, A.; van Dam, C.P. Comparison of pneumatic jets and tabs for Active Aerodynamic Load Control. *Wind Energy* **2014**, *17*, 1368–1384. [CrossRef]
2. Rumsey, C. Successes and Challenges for Flow Control Simulations (Invited). In Proceedings of the 4th Flow Control Conference, Fluid Dynamics and Co-located Conferences, AIAA 2008-4311, Seattle, WA, USA, 23–26 June 2008. [CrossRef]
3. Lengers, M. Industrial assessment of overall aircraft driven local active flow control. In Proceedings of the 29th Congress of the International Council of the Aeronautical Sciences, St. Petersburg, Russia, 7–12 September 2014.
4. Seifert, A.; Shtendel, T.; Dolgopyat, D. From lab to full scale Active Flow Control drag reduction: How to bridge the gap? *J. Wind Eng. Ind. Aerodyn.* **2015**, *147*, 262–272. [CrossRef]
5. Morgulis, N.; Seifert, A. Fluidic flow control applied for improved performance of Darrieus wind turbines. *Wind Energy* **2016**, *19*, 1585–1602. [CrossRef]
6. Vrchota, P.; Hospodář, P. Response Surface Method Application to High-Lift Configuration with Active Flow Control. *J. Aircr.* **2012**, *49*, 1796–1802. [CrossRef]
7. Galbraith, M. Numerical Simulations of a High-Lift Airfoil Employing Active Flow Control. In Proceedings of the 44th AIAA Aerospace Sciences Meeting, Reno, Nevada, 9–12 January 2006. [CrossRef]
8. Yousefi, K.; Saleh, R. Three-dimensional suction flow control and suction jet length optimization of NACA 0012 wing. *Meccanica* **2015**, *50*, 1481–1494. [CrossRef]
9. Huang, L.; Huang, P.G.; LeBeau, R.P.; Hauser, T. Numerical Study of Blowing and Suction Control Mechanism on NACA0012 Airfoil. *J. Aircr.* **2004**, *41*, 1005–1013. [CrossRef]
10. Yousefi, K.; Saleh, R. The Effects of Trailing Edge Blowing on Aerodynamic Characteristics of the NACA 0012 Airfoil and Optimization of the Blowing Slot Geometry. *J. Theor. Appl. Mech.* **2014**, *52*, 165–179. [CrossRef]
11. Rumsey, C.L.; Nishino, T. Numerical study comparing RANS and LES approaches on a circulation control airfoil. *Int. J. Heat Fluid Flow* **2011**, *32*, 847–864. [CrossRef]

12. Schatzman, D.M.; Wilson, J.; Maron, L.; Palei, V.; Seifert, A.; Arad, E. Suction and oscillatory blowing interaction with boundary layers. In Proceedings of the 53rd AIAA Aerospace Sciences Meeting, AIAA SciTech Forum, Kissimmee, FL, USA, 5–9 January 2015.

13. Lakebrink, M.T.; Mani, M.; Winkler, C. Numerical Investigation of Fluidic Oscillator Flow Control in an S-Duct Diffuser. In Proceedings of the 55th AIAA Aerospace Sciences Meeting, Grapevine, TX, USA, 9–13 January 2017.

14. Vrchota, P. Active Flow Separation Control Applied at Wing-Pylon Junction of a Wing Section in Landing Configuration. In Proceedings of the 55th AIAA Aerospace Sciences Meeting, Grapevine, TX, USA, 9–13 January 2017; p. 0491.

15. Vatsa, V.; Koklu, M.; Wygnanski, I. Numerical Simulation of Fluidic Actuators for Flow Control Applications. In Proceedings of the 6th Flow Control Conference, Fluid Dynamics and Co-located Conferences, New Orleans, LA, USA, 25–28 June 2012; AIAA 2012-3239. [CrossRef]

16. Kara, K. Numerical Study of Internal Flow Structures in a Sweeping Jet Actuator. In Proceedings of the 33rd AIAA Applied Aerodynamics Conference, Atlanta, GA, USA, 16–20 June 2015; AIAA 2015-2424.

17. Troshin, V.; Palei, V.; Shtendel, T.; Seifert, A. Characterization of the Suction and Oscillatory Blowing Actuator in Still Air. In Proceedings of the 54th Israel Annual Conference on Aerospace Sciences, Tel Aviv, Israel, 19–20 February 2014.

18. Arwatz, G.; Fono, I.; Seifert, A. Suction and oscillatory blowing actuator modeling and validation. *AIAA J.* **2008**, *46*, 1107–1117. [CrossRef]

19. Wallin, S.; Johansson, A.V. An explicit algebraic Reynolds stress model for incompressible and compressible turbulent flows. *J. Fluid Mech.* **2000**, *403*, 89–132. [CrossRef]

20. Gritskevich, M.S.; Garbaruk, A.V.; Schütze, J.; Menter, F.R. Development of DDES and IDDES Formulations for the k–ω Shear Stress Transport Model. *Flow Turbul. Combust.* **2012**, *88*, 431–449. [CrossRef]

21. Dolgopyat, D.; Seifert, A. Active Flow Control Virtual Maneuvering System Applied to Conventional Airfoil. *AIAA J.* **2019**, *57*, 72–89. [CrossRef]

22. Kim, J.; Moin, P.; Seifert, A. Large-Eddy Simulation-Based Characterization of Suction and Oscillatory Blowing Fluidic Actuator. *AIAA J.* **2017**, *55*, 2566–2579. [CrossRef]

23. Vrchota, P.; Prachař, A.; Hospodář, P.; Dolgopyat, D.; Seifert, A. Development and Verification of Simplified Boundary Condition of SaOB Actuator. In Proceedings of the 58th Israel Annual Conference on Aerospace Sciences, Tel Aviv, Israel, 14–15 March 2018.

24. Dolgopyat, D.; Seifert, A. Conventional Airfoil Active Flow Control Virtual Maneuvering System. In Proceedings of the 55th Israel Annual Conference on Aerospace Sciences, Tel Aviv, Israel, 25–26 February 2015.

25. Vrchota, P.; Prachař, A.; Hospodář, P.; Dolgopyat, D.; Seifert, A. Development of Simplified Boundary Condition of SaOB Actuator Based on High-Fidelity CFD Simulations. In Proceedings of the 52nd 3AF International Conference on Applied Aerodynamics—Progress in Flow Control, Lyon, France, 27–29 March 2017.

26. Prachař, A.; Vrchota, P. Characterization of the Suction-and-Oscillatory-Blowing actuator by the hybrid RANS-LES CFD. In Proceedings of the 2018 Flow Control Conference, AIAA AVIATION Forum, Atlanta, GA, USA, 25–29 June 2018. [CrossRef]

27. Eliasson, P. *EDGE, a Navier-Stokes Solver for Unstructured Grid*; ISTE Ltd.: London, UK, 2002; pp. 527–534.

28. Swanson, R.C.; Radespiel, R.; Turkel, E. On Some Numerical Dissipation Schemes. *J. Comput. Phys.* **1998**, *147*, 518–544. [CrossRef]

29. Zhang, X.S. *Neural Networks in Optimization*; Volume 46; Springer Science & Business Media: Berlin/Heidelberg, Germany, 2013.

30. Rasmussen, C.E.; Williams, C.K.I. *Gaussian Processes for Machine Learning (Adaptive Computation and Machine Learning)*; The MIT Press: Cambridge, MA, USA, 2005.

31. Jirasek, A. Mass flow boundary conditions for subsonic inflow and outflow boundary. *AIAA J.* **2006**, *44*, 939–947. [CrossRef]

Article

Challenges of Fully-Coupled High-Fidelity Ditching Simulations

Maximilian Müller * **, Malte Woidt, Matthias Haupt and Peter Horst**

Institute of Aircraft Design and Lightweight Structures, Technische Universität Braunschweig, Hermann–Blenk–Str. 35, D–38108 Braunschweig, Germany; m.woidt@tu-braunschweig.de (M.W.); m.haupt@tu-braunschweig.de (M.H.); p.horst@tu-braunschweig.de (P.H.)
* Correspondence: maximilian.mueller@tu-braunschweig.de; Tel.: +49-531-391-9919

Received: 5 November 2018; Accepted: 14 January 2019; Published: 22 January 2019

Abstract: An important element of the process of aircraft certification is the demonstration of the crashworthiness of the structure in the event of an emergency landing on water, also referred to as ditching. Novel numerical simulation methods, that incorporate the interaction between fluid and structure, open up a promising way to model ditching in full scale. This study focuses on two main issues of high-fidelity ditching simulations: the development of a suitable fluid-structure coupling framework and the generation of the structural model of the aircraft. The first issue is addressed by implementing a partitioned coupling approach, which combines a finite volume hydrodynamic fluid solver as well as a finite element structural solver. The developed framework is validated by means of two ditching-like experiments, which consider the drop test of a rigid cylinder and a deformable cylindrical shell. The results of the validation studies indicate that an alternative to the standard Dirichlet-Neumann partitioning approach is needed if a strong added-mass effect is present. For the full-scale simulation of aircraft ditching, structural models become more complex and have to account for damage. Due to its high localization, the damage creates large differences in model scale and usually entails severe non-linearities in the model. To address the issue of increasing computational effort for such models, the process of developing a multi-scale model for the simulation of the failure of fuselage frames is presented.

Keywords: crashworthiness; multiscale damage model; ditching simulation; fluid-structure interaction; incompressible flow

---

1. Introduction

Commercial transport aircraft often fly above open waters and, thus, have to prove the safe landing on water. This process is also referred to as ditching and is part of the certification specifications by the European Aviation Safety Agency (EASA) and the Federal Aviation Administration (FAA) (see [1,2]). Those regulations aim for minimizing the risk of injury for the occupants of the aircraft during the impact phase as well as allowing a safe evacuation in the subsequent floatation phase. This requires an accurate prediction of the structural behaviour of the aircraft during the impact phase as well the damage and its influence on the floatation capabilities.

A common practice to certify aircraft for ditching is to compare the new structural design with existing aircraft configurations [3]. As this approach impedes the development of novel, unconventional aircraft configurations, contemporary aircraft design relies on alternative methods to analyse ditching. Experimental investigations of ditching are limited to low horizontal speed (e.g., helicopters [4]) or are conducted on model scale [5,6] where they suffer from scaling effects, particularly with respect to the accurate representation of hydrodynamic effects [7]. Consequently, there is a motivation to employ numerical tools to simulate ditching. Those tools can be broadly classified into *high-fidelity* and *low-fidelity* approaches. High-fidelity methods are based upon general computational mechanics and usually employ particles or a mesh to discretize the aircraft structure and the surrounding fluid. Conversely, low-fidelity methods discretize structure and fluid in a simplified manner and are based on analytical or semi-analytical formulations (see e.g., [8,9]). Those tools are often used in the context of a one-way coupling approach, where the hydrodynamic loads are calculated by simplified models and are subsequently applied to a high-resolution finite element (FE) model. In summary, it can be stated that, due to the impracticability of full-scale tests, the limitations of model scale experiments as well as the simplifications in numerical approaches, there is a demand for advanced, high-fidelity simulation methods.

The development of such methods is an active field of current research. Many publications focus on the simulation of ditching events at zero or low horizontal speed (see e.g., [4,10,11]). This condition is normally met during the ditching of helicopters for which results of full-scale experiments are available and are used to validate simulations [4]. Conversely, the ditching of fixed-wing aircraft involves a significant horizontal speed and, therefore, hydrodynamic effects such as cavitation, ventilation and the formation of bow waves and spray jets need to be considered [3,12]. These physical effects present additional numerical challenges compared to helicopter ditching.

Furthermore, the large hydrodynamic loads arising in the impact zone usually rise and fall within milliseconds and therefore require small time step sizes in the fluid simulation [3]. Those loads result in local structural deformation and possibly failure. Moreover, failure occurring in fuselage frames usually occurs locally as well. If modelled using a high-resolution FE discretization, this may lead to large differences in model scale [13].

As the flexibility of the structure has a significant effect on the hydrodynamic loads acting on the structure during the ditching event, fluid-structure interaction (FSI) needs to be considered [14]. Due to the specific requirements of the individual solvers, monolithic coupled solvers are not practical. Instead, a partitioned coupling approach is required.

The process of ditching is accompanied by a strong added-mass effect of the fluid on the structure [15]. This gives rise to numerical instabilities in the coupled solution, particularly if fluid and structural densities are comparable, the structure is slender and incompressible flow is assumed [16–18]. Iterative (implicit) coupling methods have been widely adopted to simulate strongly coupled problems with a pronounced added-mass effect (see e.g., [19,20]). However, those methods may also suffer from stability issues and increased computational effort [18]. It is therefore crucial to employ an efficient partitioned procedure in order to make highly resolved ditching simulations feasible.

Current high-fidelity methods usually employ the FE method to simulate the structural response of the aircraft. Conversely, the fluid domain typically is discretized by using either a mesh-free approach (see e.g., [21]) or a hybrid approach. The latter combines classical Lagrangian FE methods with smoothed-particle hydrodynamics (SPH) to alleviate the problem of growing computational effort with larger fluid domains (see e.g., [3,6,22]). The advantage of such particle methods is that they eliminate the issue of mesh deformation in the fluid domain. However, cavitation is only accounted for in a simplified manner by cutting off the pressure as soon as it falls below a defined minimum value [3,23]. Hence, both phase changes and the presence of vapour are not considered, although those effects may considerably influence the hydrodynamic loads of the fluid on the structure [8,24]. This can be remedied by using the finite volume (FV) method for which cavitation models are already available and tested (see e.g., *interPhaseChangeFoam* solver in *OpenFOAM 5*).

The focus of this study is twofold: a suitable fluid-structure coupling framework is presented and validated using the two-dimensional drop test of a rigid cylinder and a deformable cylindrical shell. Furthermore, the process of developing a multiscale model, which simulates the failure of fuselage frames, is illustrated. The paper is structured accordingly: the next section introduces the numerical framework as well as an implicit fixed-point coupling scheme. In Section 3.1, this scheme is employed to simulate the validation problems mentioned above. The elastic validation problem will additionally be simulated using a simple staggered coupling scheme. Section 3.2 deals with the development of a multiscale structural model of the aircraft fuselage. The particular challenge is to accurately model structural failure while limiting the computational effort to a manageable amount. A summary along with conclusions are given in Section 4.

2. Numerical Simulation Framework

This study employs the coupling environment *ifls*, which has already been used to model various multi-physics problems (see e.g., [25–27]). *Ifls* follows a modular approach that allows to couple two or more black-box solvers in a partitioned way. The usage of the interpreted, high-level programming language *Python* facilitates a flexible and easy implementation of the desired coupling scheme. Time-critical operations are performed using compiled code as well as the *NumPy* library. Figure 1 shows the individual processing units of *ifls*. Simulation steering and solver communication are performed by the *control code* and the *co-processes*, respectively. Each of those units is run in a separate thread.

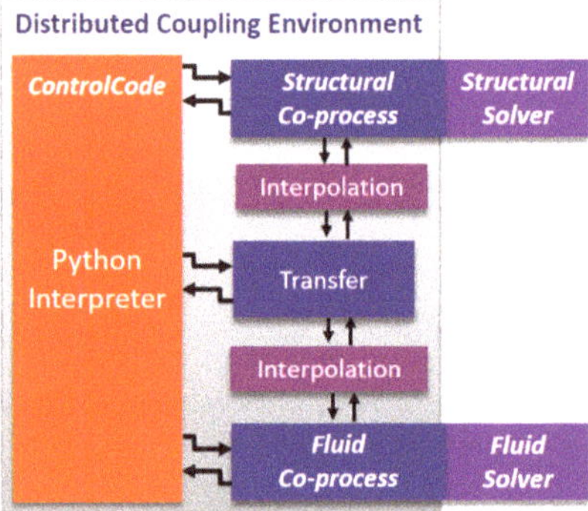

Figure 1. Coupling environment *ifls* (see [25–27]).

The control code implements the selected partitioned coupling scheme and sends control commands to the co-processes. Those commands provide direct control over the respective solvers and employ a predefined set of instructions that includes, for example, solving a time step or defining and obtaining boundary conditions. Additional instructions for sending and receiving coupling quantities enable access to the coupling data. *NumPy* is used to store those data and perform arithmetic operations, thus circumventing the speed limitations of Python. If coupling data is exchanged between different surface meshes, the send and receive instructions implicitly invoke appropriate interpolation routines, which are implemented in *converter* objects and employ methods from the *visualisation toolkit* (VTK) (The visualization toolkit is Open-Source software, which can be downloaded from https://www.vtk.org.) As the interpolation is computationally more expensive, all interpolation filters are implemented in C++. Another advantage of the usage of VTK is that it facilitates the easy visualization of the simulation results.

Each co-process implements the instruction set used by the control code where basic features common to all co-processes, for example, inter-thread communication and quantity array management, are separately implemented in a generic parent class. Communication between a co-process and the corresponding solver can be achieved in different ways. In this case, a socket-based approach is employed, which has superior performance compared to a simple file-based data exchange. Further information on the coupling environment *ifls* as well as its underlying principles can be found in Reference [28].

For the simulation of the fluid domain, the finite-volume solver *interDyMFoam*, which is part of the *OpenFOAM 5* toolbox, is employed. It solves the incompressible Navier-Stokes equations and is able to model multiphase flow by means of the *Volume of Fluid* (VOF) method. The fluid equations are solved in the *Arbitrary Lagrangian-Eulerian* (ALE) form to account for mesh movement in the convective term. The communication between *ifls* and *OpenFOAM* is achieved by using a socket-based data exchange as presented in Reference [29]. The structure is simulated with the implicit *Abaqus* solver assuming linear elastic material behaviour.

The general partitioned FSI problem can be formulated as follows. The computational domain Ω is divided into the structural domain Ω_S and the fluid domain Ω_F including their respective boundaries $\partial\Omega_S$ and $\partial\Omega_F$. Both domains share the common boundary $\Gamma_i \overset{\text{def}}{=} \partial\Omega_S \cap \partial\Omega_F$, referred to as the FSI interface, on which the coupling values are exchanged. On the fluid side of the interface, the Dirichlet boundary condition

$$v_f = v_s, \text{ on } \Gamma_i \tag{1}$$

is applied, where v_f and v_s represent the fluid and structural velocity vector. Equation (1) corresponds to a no-slip velocity boundary condition in the fluid domain.

Conversely, the fluid forces are applied as a Neumann boundary condition to the structural domain:

$$\sigma_f n = \sigma_s n, \text{ on } \Gamma_i. \tag{2}$$

The quantities σ_f and σ_s represent the fluid and structural stress tensor, respectively, whereas n denotes the unit normal vector.

Ditching exhibits typical characteristics of a strongly coupled FSI problem. For that reason, the implicit fixed-point iteration procedure summarized in Figure 2 is employed. The variables u_Γ and v_Γ contain the nodal displacements and velocities at the interface Γ_i, whereas p_Γ represents the nodal forces. The operator S describes the solution in the corresponding computational domain. The mapping of the coupling values at the FSI interface from fluid to structure and vice versa is expressed by $T_{\Gamma,\text{fs}}$ and $T_{\Gamma,\text{sf}}$. The subscripts s and f denote the structural and fluid domain. The index n represents the current time increment, whereas i stands for the current iteration.

Outer time loop (index n)	
$(\tilde{u}_{\Gamma,s})_n = (u_{\Gamma,s})_{n-1} + \Delta t \left[\frac{3}{2}(v_{\Gamma,s})_{n-1} - \frac{1}{2}(v_{\Gamma,s})_{n-2}\right]$	Calculate structural displacements using second order predictor
$(u_{\Gamma,s})_n^{i=0} = (\tilde{u}_{\Gamma,s})_n$	Set structural displacement values of the first inner iteration to predicted displacements
Inner equilibrium iterations (index i)	
$(u_{\Gamma,f})_n^i = T_{\Gamma,\text{sf}}\left((u_{\Gamma,s})_n^i\right)$	Map structural displacements to fluid mesh
$(p_{\Gamma,f})_n^i = S_f\left((u_{\Gamma,f})_n^i, (u_{\Gamma,f})_{n-1}^i\right)$	Fluid solution using previous and current displacement values
$(p_{\Gamma,s})_n^i = T_{\Gamma,\text{fs}}\left((p_{\Gamma,f})_n^i\right)$	Map fluid forces to structural mesh
$(u_{\Gamma,s})_n^i = S_s\left((p_{\Gamma,s})_n^i, (u_{\Gamma,s})_{n-1}^i\right)$	Structural solution assuming linear variation of the force values within the time step
$(u_{\Gamma,s})_n^i = (1 - \omega)(u_{\Gamma,s})_n^{i-1} + \omega \cdot (u_{\Gamma,s})_n^i$	Relaxation of current structural displacements

Figure 2. Solution procedure for time level t_n.

The fixed-point scheme is usually suitable for strongly coupled problems, although it requires several equilibrium iterations per time step. To accelerate the convergence, the *Aitken relaxation* method is used to calculate the relaxation factor ω [30]. Convergence is checked at the end of each iteration by the following displacement-based criterion:

$$r_{\Gamma,s} = \frac{\left\| (u_{\Gamma,s})_n^{i-1} - (u_{\Gamma,s})_n^i \right\|_2}{\left\| (u_{\Gamma,s})_n^i - (u_{\Gamma,s})_{n-1} \right\|_2 + C} < r_{\Gamma,s,\max} \tag{3}$$

where the constant value C ensures that the denominator does not become zero.

3. Results and Discussion

3.1. Validation Studies

3.1.1. Drop Test of Rigid Cylinder

The model problem of the drop test of a rigid cylinder is based on an experimental study by Greenhow and Lin [31]. The problem set-up as well as the boundaries of the computational domains are shown in Figure 3a. Boundary $\Gamma_{f,\mathrm{I}}$ represents the free atmosphere, whereas the boundaries $\Gamma_{f,\mathrm{II-IV}}$ are fixed walls. Table 1 summarizes the velocity and pressure boundary conditions which are valid in the fluid domain. Moreover, the interface conditions described in Section 2 are applied on the cylinder surface. The diameter of the cylinder is set to $d = 0.11$ m. The mass is chosen in such a way that the cylinder is neutrally-buoyant. The impact velocity amounts to $\left| v_{entry} \right| = 2.955$ m/s.

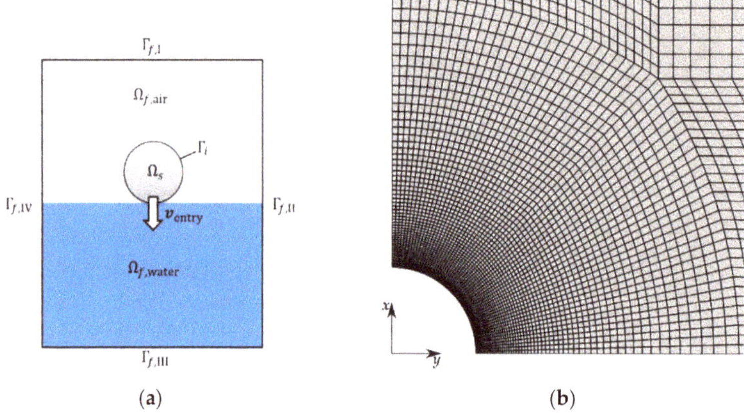

(a)	(b)

Figure 3. Drop test of rigid cylinder: (**a**) Schematic representation of drop test; (**b**) Section of the fluid mesh (mesh02).

Table 1. Boundary conditions in fluid domain.

Boundary	Pressure	Velocity
		Outflow: zero gradient in normal direction
$\Gamma_{f,\mathrm{I}}$	Constant zero pressure	Inflow: Robin condition based on flux in the patch-normal direction
$\Gamma_{f,\mathrm{II-IV}}$	Zero gradient in normal direction	Constant zero velocity

The solid is meshed with rigid surface elements where point masses are used to model the mass of the cylinder. The fluid domain is spatially discretized by a two-dimensional block-structured mesh with an O-Grid around the cylinder. A section of the initial fluid mesh *mesh02* is shown in Figure 3b. This mesh consists of 22k cells and is constructed from 90 elements along the outer edges $\Gamma_{f,\mathrm{I}}$ and $\Gamma_{f,\mathrm{III}}$

to 120 elements along $\Gamma_{f,\text{II}}$ and $\Gamma_{f,\text{IV}}$. The mesh density increases in closer proximity to the cylinder. Additionally, two h-refinement steps are performed. The resulting meshes are labelled *mesh03* and *mesh04* and contain 89k and 355k cells, respectively. For the simulation of the rigid validation case, the medium fine mesh03 is used.

Fluid and structural simulation are coupled employing the fixed-point iteration scheme described in Section 2. The displacement residual is set to $r_{\Gamma,s,\max} = 10^{-7}$ m, while a constant time step size of 2.5×10^{-5} s is chosen. In the structural simulation, the time integration is done by an implicit Newmark-beta method, whereas the fluid solver uses the implicit Euler method. Furthermore, the volume of fluid method in combination with the so-called Multi-Dimensional Limiter for Explicit Solution (MULES) interface compression scheme is employed to model the free water surface. For a detailed description of that scheme, see for example [32].

Figure 4b depicts the depth of penetration over the time and shows good agreement between simulation and experiment. However, the illustration of the different stages of water impact in Figure 5 reveals that the free water surface is not perfectly sharp at the beginning of the simulation. This indicates a strong dependence of the MULES method on the mesh geometry, which was for example also found in Reference [33]. As the simulation results match the experimental results well, it can be assumed that this effect does not have a significant influence on the overall hydrodynamic loads.

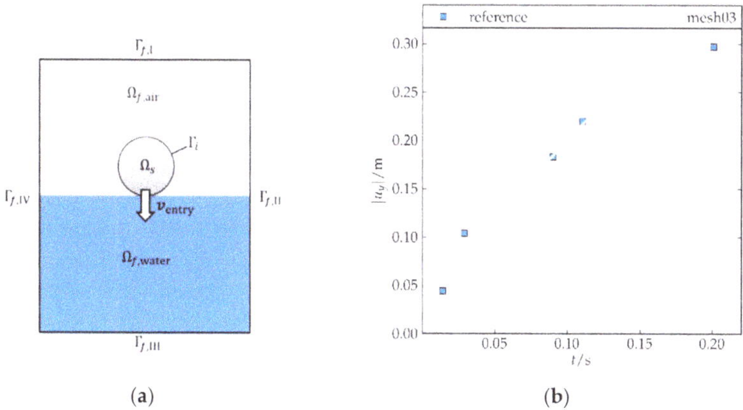

(a) (b)

Figure 4. Drop test of rigid cylinder: (**a**) Schematic representation of drop test; (**b**) Results from drop test of rigid cylinder, experimental results from [31].

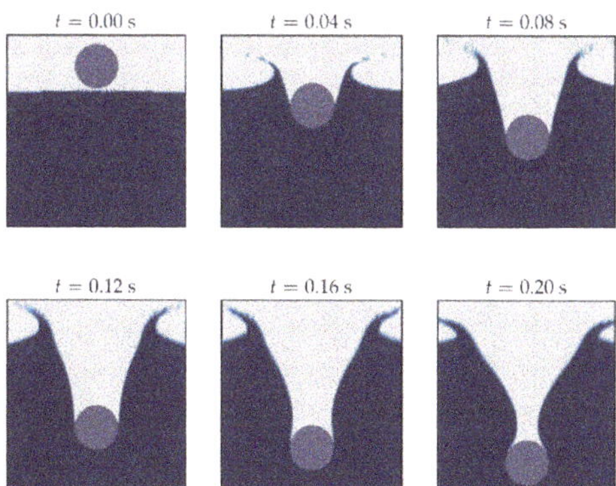

Figure 5. Stages of water impact of rigid cylinder.

3.1.2. Drop Test of Deformable Cylinder

The elastic validation case is based on [34] by Arai and Miyauchi. In their study, the impact of cylindrical shells on water is both experimentally and numerically investigated. The parameters of the second validation case are summarized in Table 2.

Table 2. Parameters of elastic validation case.

Parameters	Value
Outer cylinder diameter	$d = 0.306$ m
Shell thickness	$t = 3 \times 10^3$ m
Entry velocity	$\left| v_{\text{entry}} \right| = 3.968$ m/s
Shell density	$\rho_{s,1} = 2.70 \times 10^3 \text{kg m}^{-3}$ $\rho_{s,2} = 3.04 \times 10^3 \text{kg m}^{-3}$
Young's modulus	$E_s = 7.35 \times 10^{10}$ Pa
Poisson's ratio	$\nu_s = 0.34$

As proposed in Reference [35], the friction of the sliding system in the experiment taken into account by assuming that the water entry velocity is reduced by 10% to the value shown in Table 2. Furthermore, following a simplified approach proposed by Ionina and Korobkin in Reference [36], additional masses due to cables and connectors are considered through an increase of the shell density by 11%, that is assuming those masses are uniformly distributed. In this context, $\rho_{s,1}$ represents the density without additional mass, whereas $\rho_{s,2}$ stands for the cylindrical shell with additional mass.

Despite different dimensions, the discretization of the fluid model is equivalent to the one described in Section 3.1.1, except that all three fluid mesh configurations are investigated in the deformable validation case. The time step is set to a constant value of 1×10^{-5} s. The time integrators used in the fluid and structural subsystem are the same as for the rigid validation case. The structural mesh shown in profile in Figure 6 consists of elastic shell elements with linear form functions.

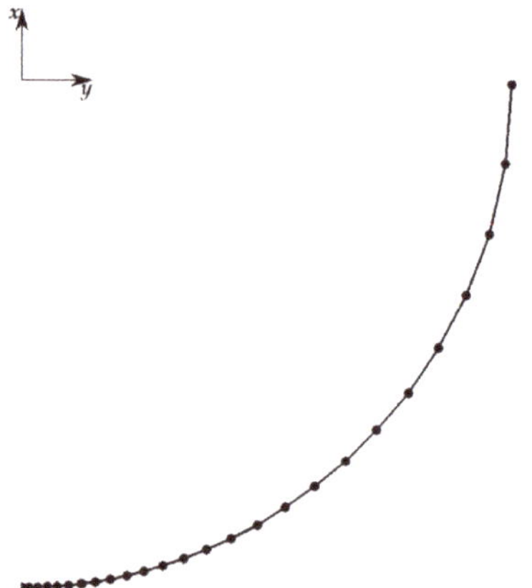

Figure 6. Section of the structural mesh of deformable cylindrical shell.

It represents the coarsest resolution and is used in combination with mesh02. For the finer configurations mesh03 and mesh04, h-refinement steps are similarly applied to the structural mesh. Again, the implicit coupling scheme described in Section 2 is used. In contrast to the rigid case, the displacement residual, which is set to $r_{\Gamma,s,max} = 10^{-10}$ m, is complemented by a maximum number of 30 iterations per time increment.

Figure 7a compares the experimental strain response at the bottom of the cylinder with the numerical values of all three mesh configurations for a structural density of $\rho_s = \rho_{s,2}$. The solution is already sufficiently converged for mesh03. For all mesh configurations, the results reveal an overprediction of the magnitude of extreme values. For the two global extreme values, the difference amounts to 13% on average for mesh02 and 20% for mesh04. One reason for this difference could be the insufficient resolution of the experimental data, which does not reproduce higher frequent oscillations in the strain response and may prevent an accurate representation of extreme values. An additional observation is that the dynamic response of the cylinder is slightly faster in the simulation.

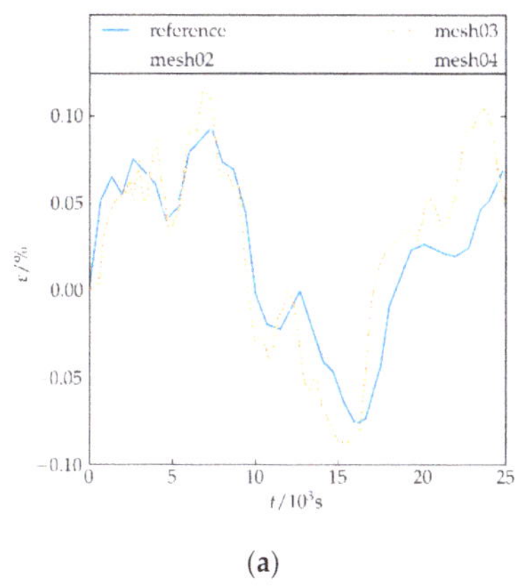
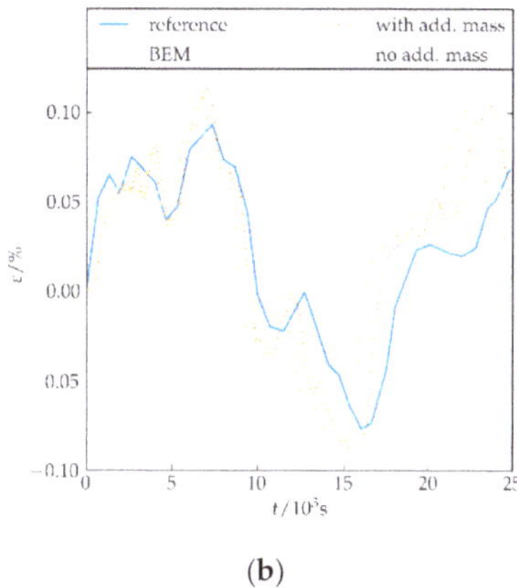

(a)   (b)

Figure 7. Strain response of deformable cylinder (reference data from [34,35]): (**a**) Comparison of different mesh configurations using $\rho_s = \rho_{s,2}$; (**b**) Comparison of different structural densities.

Similar deviations were found by Sun and Faltinsen in Reference [35] by their two-dimensional boundary element method (BEM). The corresponding results are shown in Figure 7b. The phase shift is similar to the one found in this study. However, the difference in the magnitude of extreme values is smaller. As explained by Sun and Faltinsen, the extreme values significantly depend on the force calculation at the beginning of the simulation. If those forces are calculated by the more accurate Wagner's method (see [37]), the magnitudes of the extreme values become larger.

In summary, two factors can lead to the differences observed between simulation and experiment. On the one hand, the resolution of the experimental data is relatively low. This could lead to inaccurate extreme values. On the other hand, certain assumptions are not fulfilled in reality, for instance, additional masses are not uniformly distributed within the cylinder. This may alter the dynamic behaviour of the structure.

Additionally, Figure 7b compares the strain response for the different structural densities $\rho_{s,1}$ and $\rho_{s,2}$. It becomes apparent that negligence of the additional mass leads to a less accurate approximation of the experimental data.

To assess the computational efficiency of the iterative coupling method, the number of iterations required each time increment during the moment of impact is compared for all mesh configurations in Figure 8. Although some increments reach the maximum number of 30 iterations, the average number of iterations per increment over the whole simulation is lower. It amounts to 24.5 for mesh02, 22.9 for mesh03 and 17.9 for mesh04. Anyhow, this number is still too large to directly employ such a coupling scheme for the high-fidelity simulation of a realistic aircraft structure.

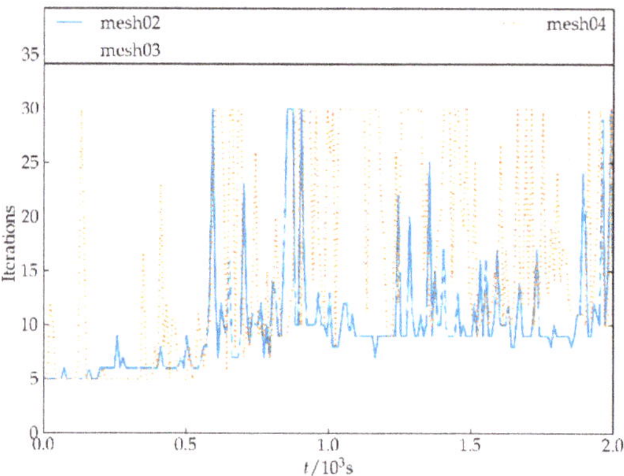

Figure 8. Iterations needed per time step during impact.

For that reason, alternative coupling procedures are needed. Also, as the modelling of structural damage is a major component of high-fidelity ditching simulations, explicit time integrations schemes should be used in the structural simulation. A simple staggered coupling scheme, which is exchanging the coupling data only once per time step, seems to be a viable solution. However, this scheme is not stable if standard Dirichlet-Neumann boundary conditions are employed and, thus, requires alternative partitioning approaches.

3.2. Structural Model

3.2.1. Global FE-Model

Preliminary ditching simulations with a rigid aircraft similar to the A350 show that an approximately 10 m long section of the rear fuselage is subjected to water loads during the impact phase of a ditching event. The relevant section is shown in Figure 9. Though only a small part of the fuselage is subjected to the hydrodynamic impact loads, the whole aircraft needs to be structurally modelled to capture the overall dynamics during impact. If structural damage occurs, it is expected to be limited to the green region.

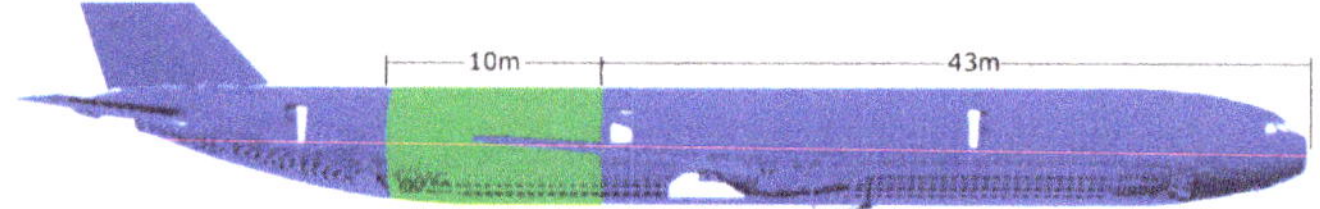

Figure 9. Relevant structural region (highlighted in green) of an A350-type aircraft.

To capture the overall aircraft dynamics outside the green region, a global FEM model (GFEM) based on the XRF-1, an Airbus large transport aircraft of the eXternal Research Forum (XRF), is used. Stringers and frames are modelled as rods and beams on a skin modelled as shells, whereas masses are defined as point masses which are assigned to whole fuselage sections. The total number of degrees of freedom in the model amounts to approximately 300k.

3.2.2. Fuselage Crash Simulations

The majority of the available methods to simulate fuselage damage under crash conditions primarily address barrel drop tests on solid ground. Two distinct numerical approaches can be found in the literature.

The first approach is to discretize the airframe as a lumped mass-spring-damper system. This leads to small models with only several hundred degrees of freedom. Examples of such an implementation are *DRI-KRASH* solver described in Reference [13] or a method by *Weiß* proposed in Reference [38]. Obvious advantages of those approaches are their simplicity of application and reduced computational effort. This allows examining many different configurations and crash scenarios. However, the parameters needed for those models have to be obtained experimentally or from experience. Furthermore, the results need to be carefully reviewed as the range of validity cannot be reliably predicted.

The other approach is the simulation on a high-fidelity level using explicit FE solvers [39]. Although this approach tends to be more accurate, experimental verification still remains an important requirement. Another major disadvantage is the high computational cost. Besides the model size, the computational effort is dependent on the physical time to be simulated as well as the refinement level in the model. Especially the latter can cause problems as the maximum time step size is directly linked to the size of the smallest element in the model. A second disadvantage is that the required degree of granularity in the model needs to be determined. In this context, the term granularity not only refers to the element size but also to the level of structural detail, for instance, if discrete rivets with failure criterions are modelled. For barrel drop tests, suitable degrees of granularity were determined by comparisons between simulations and experiments. However, the question if those results are applicable to the simulation of ditching events still remains unanswered.

If a direct simulation of damage with the explicit FE method is employed, the structural models need to be fine enough to account for relevant damage effects. To illustrate this, the simulation of a breaking frame loaded by a bending moment is shown in Figure 10b. As frames serve as circumferential stiffeners, their failure has the biggest impact on the overall stiffness of the fuselage [38]. The damage is highly localized and the damage region is magnitudes of order smaller compared to the overall aircraft dimensions.

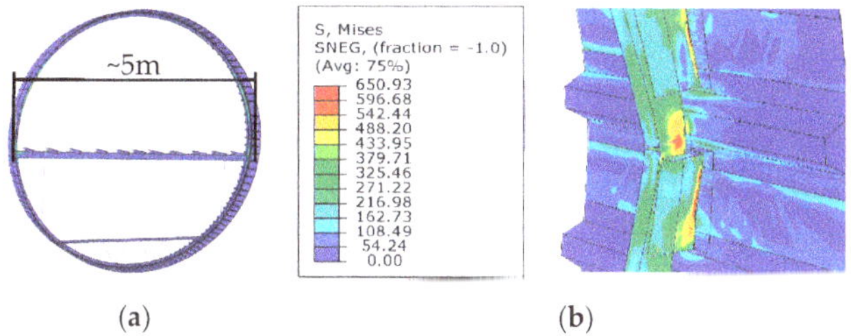

Figure 10. Structural model in detail: (**a**) Slice of the medium-detailed FE model (green region); (**b**) Damage FE model of a frame subjected to a bending moment.

Due to complex damage mechanisms, that are triggered by stability buckling in the inner flange or by plasticity effects, respectively, a very fine mesh is needed to directly simulate the damage. Consequently, a mesh size study is conducted for the case depicted in Figure 10b. The results are summarized in Figure 11, showing the bending angle over the applied momentum for different element edge lengths. The results indicate that very fine meshes are needed to get a reliable ultimate momentum. In addition, the post-failure stiffness is very dependent on the mesh size, as many damage effects are very small in scale.

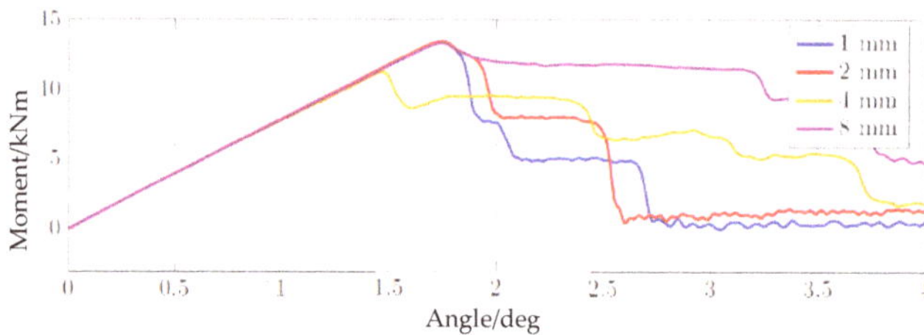

Figure 11. Results of the model in Figure 10b with different element side lengths.

The complete ditching event is expected to have a duration of several seconds and, as the computation effort grows with the simulated time, an FE model that accounts for structural damage within the region relevant for ditching by using a sufficiently fine mesh while simultaneously capturing the whole aircraft is not feasible. Therefore, the authors suggest using a multiscale model for the structural simulation.

3.2.3. Multiscale Model for Ditching Simulation

To model the section of the aircraft relevant for ditching, a combination of the two methods outlined in Section 3.2.2 is used. The green section shown in Figure 9 is modelled using an FE discretization with an element size between 15 and 20 mm. The water loads are applied to the fuselage skin. To allow the correct simulation of the load paths within the model, stringers, clips and frames are modelled as shells. The resulting model has approximately 6M degrees of freedom and, therefore, can be handled by commercial explicit FE solvers. The blue region in Figure 9 resembles the GFEM-model. Both parts of the model are connected by multi-point-constraints.

In addition to the refined mesh, a better approximation of the mass distribution is needed as inertia forces are important for the loading state during highly dynamic events. The structural masses of the refined region are directly incorporated by material densities while passenger and cargo masses are separately modelled. Afterwards, the difference in mass to the original GFEM section is evenly smeared over all elements of the model. In this way, the original mass as well as the overall mass properties of the aircraft, which are essential for the computation of the correct trajectory during ditching, are preserved. Although the moments of inertia in the green section are not preserved exactly, their influence on the overall moment of inertia is relatively small.

3.2.4. Modelling Damage

At this stage, the structural model is still too coarse to accurately account for damage effects as discussed in Section 3.2.2. Most parts of the structure are subjected to characteristic loading states. The stringers in combination with the skin are for example primarily loaded by longitudinal stresses introduced by bending of the fuselage, whereas the frames keep the fuselage cross sections in shape and are mainly loaded by bending moments and hoop stresses. The clips transfer the loads from the skin to the frame and are therefore subjected to shear stresses. These observations can be used to introduce reduced order models that can capture damage effects using a relatively coarse FE discretization. This approach is called Kinematics Model and was described by Waimer et al. in Reference [13]. For instance, the damage of a frame can be modelled as a non-linear hinge element with the centre of rotation located at the skin. The hinge is closed for uncritical loads and starts to open as soon as the frame starts to break. The moment to opening angle curve is obtained either from experimental results or, as done in his case, by using the results of a detailed FE simulation as shown for example in Figure 11. Due to the highly localized damage, its overall effect on the structural behaviour can be captured by such approaches with sufficient accuracy.

3.2.5. Creating the Multiscale Model

To create a reduced order damage model without having access to full scale ditching experiments, it is necessary to identify locations of the structure where damage might occur as well as the characteristic loading states of the corresponding parts. In addition to the frames, the vertical struts are considered as well as they have a significant influence on the damage characteristics [38]. For the creation of the multiscale model, the iterative process shown in Figure 12 is proposed.

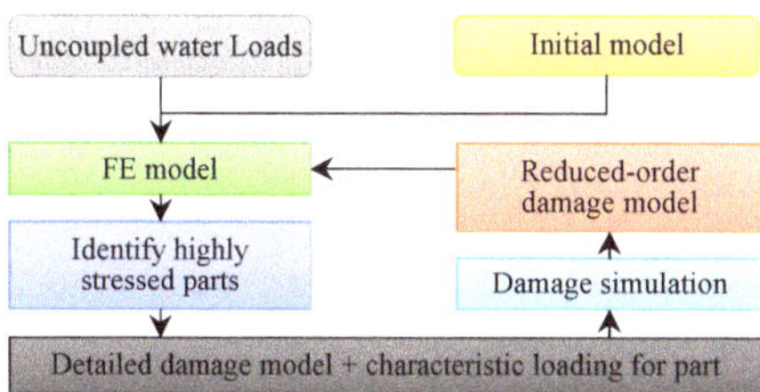

Figure 12. The iterative process of model creation.

The initial FE model is loaded by water loads obtained from a ditching simulation of a rigid structure with similar outer geometry. Highly stressed regions are identified and the characteristic loading states are extracted. For the identified regions, FE damage models with a high level of granularity are created and loaded with their respective characteristic loading states. Based on the results of the detailed simulations, appropriate surrogate models, that are able to reproduce the kinematics of the damage, are chosen and provided with the calculated load to deformation curve. The surrogate models are then introduced into the fuselage FE model. The procedure is repeated as the occurrence of damage may change the load path of the structure. If no further critical locations are found the iteration process is stopped.

4. Conclusions and Outlook

In this study, a fluid-structure coupling framework suitable for ditching simulations is presented and validated. The partitioning of the computational domain is done by the standard Dirichlet-Neumann approach. It is shown that this framework works well for two-dimensional cases with rigid and deformable structures if an implicit fixed-point iteration scheme is used. However, in the deformable case, this scheme proves to be inefficient for the iterative coupling scheme and even becomes unstable if an explicit structural solver is used together with a simple staggered coupling approach. Those results indicate that carefully designed coupling procedures need to be developed to increase computational efficiency and facilitate the usage of explicit structural solvers in the simulation of ditching problems. In this context, a simple staggered coupling scheme would be favourable but certainly requires partitioning approaches other than standard Dirichlet-Neumann boundary conditions. Possible options include operator-splitting techniques (e.g., [40]), mixed boundary conditions (e.g., [41,42]) or the weak introduction of coupling conditions (e.g., [43]). For full-scale ditching simulations, which include the failure of the structure, the time scales of fluid and structural simulation are expected to differ significantly. Hence, sub-cycling of the structural solver will likely be necessary to keep the computational effort at a manageable amount.

Furthermore, as soon as horizontal speed is introduced and mesh deformation becomes an issue, the overset grid technique is a viable option to allow large movements of the fluid-structure boundary while at the same time preserving the quality of the fluid mesh [44].

The second key point of this study pertains the development of a multiscale structural model of the aircraft fuselage. It is shown that a structural model that on the one hand can capture the overall dynamics of the aircraft and, on the other hand, directly simulate structural damage is not feasible. Due to severe nonlinearities in the structure, explicit time integration is needed. However, simply refining damage relevant regions would lead to impracticable short time steps and very poor overall computational efficiency. A method is presented how the highly localized damage behaviour of aeronautical structures can be incorporated in full-size aircraft models while retaining a model size that can be handled by today's computers. This is done by reduced order models that account for the correct kinematics of structural damage and the combination of a GFEM model with a locally refined section of interest. An iterative process is proposed how relevant regions, where reduced order damage models should be placed, can be identified.

Author Contributions: Methodology, M.M. and M.W.; Project administration, M.H. and P.H.; Software, M.M., M.W. and M.H.; Validation, M.M.; Writing–Original draft, M.M. and M.W.

Funding: The research leading to these results has received funding from the European Union's Horizon 2020 research and innovation programme under grant agreement No 724139.

Acknowledgments: We acknowledge support by the German Research Foundation and the Open Access Publication Funds of the Technische Universität Braunschweig.

Conflicts of Interest: The authors declare no conflict of interest.

References

1. European Aviation Safety Agency. Certification Specifications and Acceptable Means of Compliance for Large Aeroplanes CS-25, Book 1—Certification Specifications, Subpart C, Section CS 25.801—Ditching: CS 25.801. Amendment 22. Available online: https://www.easa.europa.eu/sites/default/files/dfu/CS-25%20Amendment%2022.pdf (accessed on 11 January 2019).
2. Federal Aviation Administration (FAA), Department of Transportation. Code of Federal Regulations, Title 14, Chapter I, Subchapter C, Part 25, Subpart D, Section 25.801—Ditching: 14 CFR 25.801. Available online: https://www.law.cornell.edu/cfr/text/14/25.801 (accessed on 11 January 2019).
3. Siemann, M.H.; Langrand, B. Coupled fluid-structure computational methods for aircraft ditching simulations: Comparison of ALE-FE and SPH-FE approaches. *Comput. Struct.* **2017**, *188*, 95–108. [CrossRef]
4. Pentecôte, N.; Vigliotti, A. Crashworthiness of helicopters on water: Test and simulation of a full-scale WG30 impacting on water. *Int. J. Crashworth.* **2003**, *8*, 559–572. [CrossRef]
5. Wernsdorfer, T.; Keller, K.; Climent, H. *CN-235-300M Deepwater—Subscale Model, Ditching and Floatation Tests Plan, Issue A*; EASA-CASA: Madrid, Spain, 2003.
6. Climent, H.; Benitez, L.; Rosich, F.; Rueda, F.; Pentecote, N. Aircraft Ditching Numerical Simulation. In Proceedings of the 25th International Congress of the Aeronautical Sciences, Hamburg, Germany, 3–8 September 2006.
7. Keller, A.P. Cavitation Scale Effects—Empirically Found Relations and the Correlation of Cavitation Number and Hydrodynamic Coefficients. In Proceedings of the CAV 2001: Fourth International Symposium on Cavitation, California Institute of Technology, Pasadena, CA, USA, 20–23 June 2001.
8. Lindenau, O.; Busch, C.; Streckwall, H.; Bensch, L.; Rung, T. Validation of a transport aircraft ditching simulation tool. In Proceedings of the 6th International KRASH Users' Seminar (IKUS6), Stuttgart, Germany, 15–17 June 2009.
9. Halbout, S.; Vagnot, A.; Markmiller, J. Modelling of water impact during ditching event. Available online: https://dspace-erf.nlr.nl/xmlui/bitstream/handle/20.500.11881/3673/ERF2016_47_S.%20HALBOUT%20_P.pdf?sequence=1 (accessed on 11 January 2019).
10. Anghileri, M.; Castelletti, L.-M.L.; Francesconi, E.; Milanese, A.; Pittofrati, M. Survey of numerical approaches to analyse the behavior of a composite skin panel during a water impact. *Int. J. Impact Eng.* **2014**, *63*, 43–51. [CrossRef]
11. Bisagni, C.; Pigazzini, M.S. Modelling strategies for numerical simulation of aircraft ditching. *Int. J. Crashworth.* **2018**, *23*, 377–394. [CrossRef]
12. Iafrati, A.; Grizzi, S.; Siemann, M.H.; Benítez Montañés, L. High-speed ditching of a flat plate: Experimental data and uncertainty assessment. *J. Fluids Struct.* **2015**, *55*, 501–525. [CrossRef]

13. Waimer, M.; Kohlgrüber, D.; Hachenberg, D.; Voggenreiter, H. The Kinematics Model—A Numerical Method for the Development of a Crashworthy Composite Fuselage Design of Transport Aircraft. In Proceedings of the Sixth Triennial International Aircraft Fire and Cabin Safety Research Conference, Atlantic City, NJ, USA, 25–28 October 2010.

14. Reinhard, M.; Korobkin, A.A.; Cooker, M.J. Water entry of a flat elastic plate at high horizontal speed. *J. Fluid Mech.* **2013**, *724*, 123–153. [CrossRef]

15. Newman, J.M. *Marine Hydrodynamics*; MIT Press: Cambridge, MA, USA, 1977.

16. Le Tallec, P.; Mouro, J. Fluid structure interaction with large structural displacements. *Comput. Methods Appl. Mech. Eng.* **2001**, *190*, 3039–3067. [CrossRef]

17. Causin, P.; Gerbeau, J.F.; Nobile, F. Added-mass effect in the design of partitioned algorithms for fluid–structure problems. *Comput. Methods Appl. Mech. Eng.* **2005**, *194*, 4506–4527. [CrossRef]

18. Förster, C.; Wall, W.A.; Ramm, E. Artificial added mass instabilities in sequential staggered coupling of nonlinear structures and incompressible viscous flows. *Comput. Methods Appl. Mech. Eng.* **2007**, *196*, 1278–1293. [CrossRef]

19. Küttler, U.; Wall, W.A. Fixed-point fluid–structure interaction solvers with dynamic relaxation. *Comput. Mech.* **2008**, *43*, 61–72. [CrossRef]

20. Küttler, U.; Gee, M.; Förster, C.; Comerford, A.; Wall, W.A. Coupling strategies for biomedical fluid–structure interaction problems. *Int. J. Numer. Methods Biomech. Eng.* **2010**, *26*, 305–321. [CrossRef]

21. Ortiz, R.; Charles, J.L.; Sobry, J.F. Structural loading of a complete aircraft under realistic crash conditions: Generation of a load database for passenger safety and innovative design. In Proceedings of the 24th International Congress of the Aeronautical Sciences, Yokohama, Japan, 29 August–3 September 2004.

22. Kohlgruber, D.; Vigliotti, A.; Weissberg, V.; Bartosch, H. Numerical simulation of a composite helicopter sub-floor structure subjected to water impact. In Proceedings of the 60th Annual Forum of the American Helicopter Society, Baltimore, MD, USA, 7–10 June 2004; pp. 7–10.

23. Groenenboom, P.H.; Campbell, J.; Benítez Montañés, L.; Siemann, M.H. Innovative SPH methods for aircraft ditching. In Proceedings of the 11th WCCM/5th ECCM, Barcelona, Spain, 20–25 July 2014.

24. Smith, A.G.; Warren, C.H.E.; Wright, D.F. *Investigations of the Behaviour of Aircraft When Making a Forced Landing on Water (Ditching)*; Her Majesty's Stationery Office: London, UK, 1957.

25. Kowollik, D.; Tini, V.; Reese, S.; Haupt, M. 3D fluid-structure interaction analysis of a typical liquid rocket engine cycle based on a novel viscoplastic damage model. *Int. J. Numer. Methods Eng.* **2013**, *94*, 1165–1190. [CrossRef]

26. Lindhorst, K.; Haupt, M.C.; Horst, P. Reduced-order modelling of non-linear, transient aerodynamics of the HIRENASD wing. *Aeronaut. J.* **2016**, *120*, 601–626. [CrossRef]

27. Sommerwerk, K.; Haupt, M.C. Design analysis and sizing of a circulation controlled CFRP wing with Coandă flaps via CFD–CSM coupling. *CEAS Aeronaut. J.* **2014**, *5*, 95–108. [CrossRef]

28. Haupt, M.; Niesner, R.; Unger, R.; Horst, P. Computational Aero-Structural Coupling for Hypersonic Applications. In Proceedings of the 9th AIAA/ASME Joint Thermophysics and Heat Transfer Conference, San Francisco, CA, USA, 5–8 June 2006; American Institute of Aeronautics and Astronautics: Reston, VA, USA, 2006; p. 95.

29. Müller, M.; Haupt, M.; Horst, P. Socket-Based Coupling of OpenFOAM and Abaqus to Simulate Vertical Water Entry of Rigid and Deformable Structures. In Proceedings of the 6th ESI OpenFOAM User Conference 2018, Hamburg, Germany, 23–25 October 2018.

30. Mok, D.P.; Wall, W.A.; Ramm, E. Accelerated iterative substructuring schemes for instationary fluid-structure interaction. *Comput. Fluid Solid Mech.* **2001**, *2*, 1325–1328.

31. Greenhow, M.; Lin, W.M. *Nonlinear-Free Surface Effects: Experiments and Theory*; Massachusetts Institute of Technology Cambridge: Cambridge, MA, USA, 1983.

32. Deshpande, S.S.; Anumolu, L.; Trujillo, M.F. Evaluating the performance of the two-phase flow solver interFoam. *Comput. Sci. Discov.* **2012**, *5*, 14016. [CrossRef]

33. Roenby, J.; Bredmose, H.; Jasak, H. A computational method for sharp interface advection. *R. Soc. Open Sci.* **2016**, *3*, 160405. [CrossRef]

34. Arai, M.; Miyauchi, T. Numerical Simulation of the Water Impact on Cylindrical Shells Considering Fluid-Structure Interaction. *J. Soc. Nav. Archit. Jpn.* **1997**, *1997*, 827–835. [CrossRef]

35. Sun, H.; Faltinsen, O.M. Water impact of horizontal circular cylinders and cylindrical shells. *Appl. Ocean Res.* **2006**, *28*, 299–311. [CrossRef]
36. Ionina, M.F.; Korobkin, A.A. Water Impact on Cylindrical Shell. In Proceedings of the 14th International Workshop on Water Waves and Floating Bodies, Port Huron, MI, USA, 11–14 April 1999.
37. Wagner, H. Über Stoss-und Gleitvorgänge an der Oberfläche von Flüssigkeiten. *Z. Angew. Math. Mech.* **1932**, *12*, 193–235. [CrossRef]
38. Weiß, L. Initial Design of Crashworthy Composite Structures. Ph.D. Thesis, Technischen Universität Carolo-Wilhelmina, Braunschweig, Germany, 2016.
39. Byar, A. A Crashworthiness Study of a Boeing 737 Fuselage Section. Ph.D. Thesis, Drexel University, Philadelphia, PA, USA, 2003.
40. Guidoboni, G.; Glowinski, R.; Cavallini, N.; Canic, S. Stable loosely-coupled-type algorithm for fluid–structure interaction in blood flow. *J. Comput. Phys.* **2009**, *228*, 6916–6937. [CrossRef]
41. Li, L.; Henshaw, W.D.; Banks, J.W.; Schwendeman, D.W.; Main, A. A stable partitioned FSI algorithm for incompressible flow and deforming beams. *J. Comput. Phys.* **2016**, *312*, 272–306. [CrossRef]
42. Badia, S.; Quaini, A.; Quarteroni, A. Modular vs. non-modular preconditioners for fluid–structure systems with large added-mass effect. *Comput. Methods Appl. Mech. Eng.* **2008**, *197*, 4216–4232. [CrossRef]
43. Burman, E.; Fernández, M.A. Explicit strategies for incompressible fluid-structure interaction problems: Nitsche type mortaring versus Robin-Robin coupling. *Int. J. Numer. Methods Eng.* **2014**, *97*, 739–758. [CrossRef]
44. Steger, J.L.; Dougherty, F.C.; Benek, J.A. A chimera grid scheme. [Multiple overset body-conforming mesh system for finite difference adaptation to complex aircraft configurations]. Advances in grid generation. In Proceedings of the Applied Mechanics, Bioengineering, and Fluids Engineering Conference, Houston, TX, USA, 20–22 June 1983; (A84-11576 02-64). American Society of Mechanical Engineers: New York, NY, USA, 1983; pp. 59–69.

Article

Comparative Environmental and Cost Analysis of Alternative Production Scenarios Associated with a Helicopter's Canopy

Christos V. Katsiropoulos *,[†] , **Andreas Loukopoulos** [†] and **Spiros G. Pantelakis** [†]

Laboratory of Technology & Strength of Materials, Department of Mechanical Engineering & Aeronautics, University of Patras, Panepistimioupolis Rion, 26500 Patras, Greece; andreasloukopoulos@upnet.gr (A.L.); pantelak@mech.upatras.gr (S.G.P.)
* Correspondence: xkatsiro@mech.upatras.gr; Tel.: +30-261-096-9498
† These authors contributed equally to this work.

Received: 9 November 2018; Accepted: 28 December 2018; Published: 3 January 2019

Abstract: In the present work the carbon footprint and the financial viability of different materials, manufacturing scenarios, as well as recycling scenarios, associated with the production of aeronautical structural components are assessed. The materials considered were carbon fiber reinforced epoxy and carbon fiber reinforced PEEK (polyetheretherketone). The manufacturing techniques compared were the autoclave, resin transfer molding (RTM) and cold diaphragm forming (CDF). The recycling scenarios included mechanical recycling and pyrolysis. For this purpose, Life Cycle Analysis (LCA) and Life Cycle Costing (LCC) models were developed and implemented for the case of a helicopter's canopy production. The results of the study pointed out that producing the canopy by using carbon fiber reinforced thermosetting composites and involving RTM as the manufacturing process is the optimal route both in terms of environmental and financial efficiency. The environmental and financial efficiency of the scenarios including thermoplastic composites as the material of choice is impaired from both the high embodied energy and raw material cost of PEEK. The scenarios investigated do not account for potential benefits arising from the recyclability and the improved reusability of thermoplastic matrices as compared to thermosetting ones. This underlines the need for a holistic aircraft structural optimization approach including not only performance and weight but also cost and environmental criteria.

Keywords: life cycle analysis; cost analysis; autoclave; resin transfer molding; cold diaphragm forming; composite materials; aeronautic component

---

1. Introduction

Carbon fiber reinforced polymer composites have been widely adopted from the aeronautical industry (e.g., 52% in Airbus A350, 50% in Boeing B787 etc.) in primary as well as secondary structures, as well as in many other similar lightweight structures, which leads to decreased fuel consumption. On the other hand, it is worth mentioning that the selection process of the appropriate material for an aircraft structural application is a vital step, made at the early design stages, and has to fulfill a number of currently applied structural optimization criteria (weight reduction, cost minimization, efficient mechanical properties, etc.). In parallel, environmental considerations should, nowadays, represent a significant parameter both for selecting a material, as well as for designing an aircraft's structure and manufacturing it.

The mainstream composite material type for aeronautical applications is thermosetting composites, mainly carbon fiber reinforced epoxies, because of their low processing temperatures, low viscosity, superior adhesion to fibers, fatigue strength, etc. However, issues associated with their long curing cycles which lead to low production rates combined with growing environmental concerns associated with their end-of-life treatment, as well as the adoption of stricter environmental policies, have turned the attention of the aeronautical industry to alternative solutions.

On the other hand, thermoplastic composites have the ability to melt under heat and harden after cooling creating a new shape. This process can be repeated multiple times. This characteristic offers possibilities for adopting faster composite processing techniques with reduced processing time. Moreover, thermoplastic composites exhibit superior impact and chemical resistance, unlimited shelf-life, and the ability to assemble sub-structures by welding, which leads to weight reduction and recyclability; properties that thermosetting composites are unable to provide. On the downside, the higher processing temperatures and pressures needed for processing these materials, which leads to both increased energy consumption and tooling cost, are key barriers for their wide-scale use for the construction of primary structures by the aeronautical industry. Additionally, the severe deterioration of the mechanical properties of the materials during the recycling process poses a burden for exploiting their recyclability.

As far as the manufacturing techniques are concerned, out of autoclave techniques demonstrate high potential for significant financial and environmental benefits as compared to the conventional autoclave process, although autoclaving is expected to lead to higher product quality. However, concerning the recycling processes, the recyclability of the material is still defined on the basis of commonly used recycling methods, including pyrolysis and landfill, which might exhibit lower energy demands. However, they lead to products with a low or without reusability potential. These conflicting parameters underline the necessity for developing tools and concepts allowing the simultaneous optimization of a product with regard to quality, cost and environmental impact.

The available works on this topic are few and limited to either the environmental footprint or the cost quantification of a product. In this context, several studies have performed either Life Cycle Analysis (LCA) of carbon fiber reinforced polymers, e.g., [1–7] or Life Cycle Costing (LCC) analysis, e.g., References [8–11] of carbon fiber reinforced polymers. An LCA study was conducted by Timmis et al. [1] to quantify the environmental impact derived from the use of composite materials in an aircraft's fuselage used in place of other structural materials (e.g., aluminum alloys). The results highlighted that the overall environmental impact which occurred from the adoption of composite materials is positive compared with traditional materials (e.g., aluminum). In Reference [2] an overview of currently used composite materials for aviation, as well as possible bio-based and recycled substitution materials, with the focus on their ecological properties is presented. Apart from conventional materials, several types of novel materials were considered to reduce ecological impacts compared to the state-of-the-art, such as bio-based thermoset resins (epoxy, furan), bio-based fibers (flax, ramie) and recycled carbon fibers. Moreover, Duflou et al. [3] and Song et al. [4] quantified the environmental footprint of composites when they replaced steel, which is of limited use in aircraft structural applications (e.g., 7% in A350), but of great importance in automotive applications. Both latter studies demonstrated that composite materials outperformed steel due to the weight savings that they offer during the in-use phase.

Furthermore, a number of studies are dedicated to the cost efficiency assessment of the composite materials, e.g., References [8–11]. In Reference [8], an overview of the extensive field of cost estimation for aerospace composite production, describing the basic methods of how to perform cost estimation and introducing some of the existing LCC models, is made. In addition, in Reference [10], an LCC procedure is developed using cost equations to accurately estimate the recurring cost to manufacture of an aircraft control surface (leading edge flap) applicable to resin transfer molding (RTM) and vacuum-assisted resin transfer molding (VARTM). A novel concept for the optimization of manufacturing processes of composite material components with regard to product's quality and cost is introduced in Reference [11] and applied for the case of thermoplastic composite helicopter canopies produced by 'cold' diaphragm forming (CDF) process.

On the other hand, a limited number of existing works refer to combined LCA and LCC analyses [12–20]; the few available works are rarely related to composites. In Reference [13] the application of an LCA/LCC integrated model is described for the comparison of an AGL (Anti-Glare Lamellae) currently manufactured from virgin HDPE (High Density PolyEthylene) with an alternative one made with recycled HDPE. The obtained results show that neither the current nor the new AGL depict the best environmental performance in all impact categories. Nevertheless, a clear overall environmental and economic advantage in replacing virgin HDPE with recycled HDPE was exhibited. Reference [14] combines LCC and LCA in the case of a residential district energy system area in Finland aiming to identify the actual technologies that could provide the highest sustainable viability and assesses the emissions and relative mitigation potentials associated with the different technologies. Furthermore, in Reference [17] an LCA study, as well as an LCC analysis, were carried out for a refractory brick production company, and in Reference [18] the evaluation of the process-based cost and environmental footprint profile of green composite under a twofold assessment is considered.

In Reference [20] a combined environmental and cost assessment dedicated to composite materials is performed. However, it is on an automotive application (steel vehicle bulkhead). In this study, the economic and environmental effects of substituting steel for lighter weight alternatives with the focus on composite materials was presented; thus, four material scenarios, as well as automated preforming technology, combined with reaction injection molding were chosen. Manufacturing and life-cycle costs were derived from a technical cost model, and the environmental performance of each scenario was then quantified using LCA according to ISO guidelines. Therefore, there is not any available work in the open literature, combining the comparison of a number of processes used in aeronautics with regards both to LCA and LCC studies for a composite aeronautic product, either thermoplastic or thermoset.

This work aims to make a comparative analysis in terms of carbon emissions and financial efficiency for different material, manufacturing, and recycling scenarios associated with the production of an aeronautical structural component; the canopy of the EUROCOPTER EC Twin Star helicopter described in Reference [11], Figure 1. In addition, the present work introduces a new holistic approach which accounts for not only cost and environmental footprint but also for additional features, such as the quality of the product as well as reusability and recyclability aspects. In such an approach, these features would be taken into account at the design phase of a product.

Figure 1. The canopy described in Refference [4] (1:3 scale prototype of the real part).

2. Methods

2.1. Life Cycle Analysis

Life cycle analysis is a standardized technique (ISO 14040 2006 [21]) for assessing the environmental performance of a product or a process at various points in their life cycle, from raw material extraction to disposal or recycling. The stages for carrying out a Life Cycle Analysis are the following: Goal and scope definition, inventory analysis, impact assessment, and interpretation of the results.

Based on this concept, an LCA model was developed for evaluating the environmental footprint in terms of carbon emissions of the canopy of the Twin Star Helicopter, which is the case study under investigation, acting as the functional unit of the system. The stages taken into consideration for the present LCA were: carbon fiber production, epoxy (TS) and PEEK (PolyEtherEtherKetone)-(TP) production, manufacturing (autoclave, resin transfer molding (RTM), cold diaphragm forming (CDF)) and recycling (mechanical recycling and pyrolysis).

Epoxy resins are the most commonly used matrix material for composite material aircraft structural applications. In this context, toughened epoxies have found widespread application. However, their use in high-performance applications is limited by low service temperature that is adversely affected by moisture content, loading, and by the use of toughening agents. In general, the maximum use temperature for advanced epoxies is limited to 150 °C to 180 °C. Other disadvantages include brittleness and moisture absorption that can lower use temperature as mentioned above. However, epoxy resins provide many attractive features, including good handling properties, processability, and low cost.

On the other hand, PEEK (PolyEtherEtherKetone) is commonly used for the matrix of TP prepregs containing mainly carbon fibers and is selected as the most prevalent TP resin used in aeronautics. PEEK has been reported to be capable of withstanding continuous operating temperature up to 260 °C in low-stress operations and 120 °C in aerospace structural applications. Moreover, PEEK has good resistance to hydrolysis, corrosion, chemical, and radiation exposure. It provides high thermal stability, a low coefficient of expansion, good abrasion resistance, low smoke and toxic gas emission, and excellent stiffness. Last, but not least, an essential advantage is recyclability which, however, requires high energy consumption.

In the present work, a Bisphenol-A epoxy-based vinyl ester resin was selected as the TS representative of toughened epoxies and PolyEtherEtherKetone (PEEK) resin as a TP competitor, both reinforced with carbon fibers. These systems were chosen to compare the environmental impact and cost of a commonly used non-recyclable material (epoxy composite) with a recyclable material (PEEK composite).

In regard to the processes under investigation, autoclave and RTM, as a representative of autoclave processes, were selected as they are the most commonly used processes to produce structural aeronautical components. They are capable of producing parts of acceptable quality according to the aeronautics standards. Cold diaphragm forming (CDF), has a low cost and seems to be promising for aeronautical applications. Therefore, the above mentioned three processes were selected for our investigation. CDF was considered only for the thermoplastic composite, as it cannot be used to process thermosetting matrices, whereas autoclave and RTM can be used for both. Moreover, pyrolysis was considered only for the scenarios involving thermosetting composite as the material of choice whereas mechanical recycling was considered for both thermoplastic and thermosetting composites. The processes under investigation were assumed to be all electric.

For each stage, the total energy demands, as well as the Global Warming Potential-100 (GWP_{100}) for a period of 100 years, were estimated. GWP_{100} was selected as the most critical environmental impact category commonly used among LCA studies for evaluating the carbon footprint and is suitable for this study as electricity generation produces mainly CO_2 emissions. The system boundaries are shown in Figure 2. The total energy consumption was calculated by multiplying the mass of each material with the energy intensity of each process (Equation (1)), which was derived from literature [5,22–25] and is shown in Table 1. Since the processing temperature of PEEK is almost three times higher than the processing temperature of epoxy, the energy demands at the manufacturing stage were assumed to be three times higher as well. The GWP_{100} was calculated by multiplying the m_{CO2eq}, which is the mass of CO_2 emission that is produced from the consumption of 1 kWh of electricity, with the total energy demands of each process (Equation (2)). The m_{CO2eq} was considered equal to 0.34 kg CO_{2eq}/kWh [26].

$$E_i = m_i e_i \qquad (1)$$

$$GWP_{100} = E_i \frac{m_{CO_2eq}}{kWh} \qquad (2)$$

where E_i is the total energy of each process in kWh, m_i is the total mass in kg of each material involved, e_i is the energy intensity for each process in kWh/kg, GWP_{100} is the Global Warming Potential for a period of 100 years in kg, m_{CO2eq} is the mass in kg of CO_2 emission that is produced from the consumption of 1 kWh of electricity.

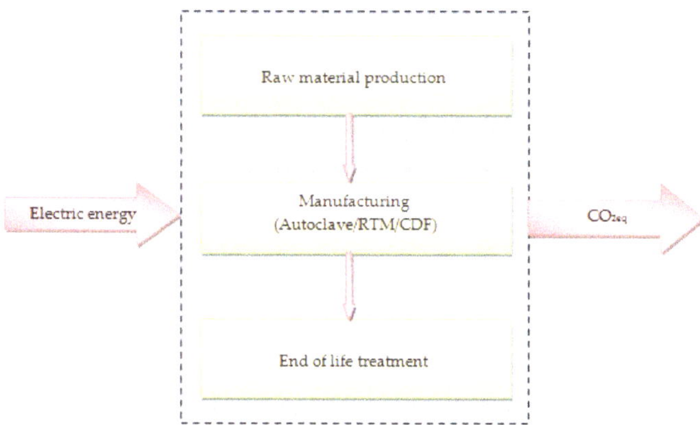

Figure 2. System boundaries.

Table 1. Energy intensity of each process.

Process	Energy Intensity (kWh/kg)
Carbon fiber production	155 [22]
Epoxy production	21.11 [5,23]
PEEK production	77.78 [24]
Autoclave TS	4.97 [22]
Autoclave TP	14.9
RTM TS	2.86 [22]
RTM TP	8.58
CDF	9 [11]
Mechanical recycling	0.075 [25]
Pyrolysis	6.67 [26]

2.2. Life Cycle Costing

The Life Cycle Cost analysis is a suitable tool for determining financial trade-offs arising from a product or a process. In this study, an LCC model was developed based on the principles of the Activity Based Costing method (ABC). ABC is an accounting method implementing cost estimation mathematical relationships (CERs) for expressing the cost as a function of one or more independent variables that take into account the geometrical features of a material or a product, such as the perimeter (PAP), surface (PAA), length (L) shape complexity (cmp), and mass (WP). In the performed analysis, costs associated with labor, material, and energy were calculated. The performed cost analysis also accounts for recycling cost which was assumed equal to the energy cost of the recycling process. CERs were either formulated or adopted from Reference [11].

In the framework of the LCC analysis, the cost of 1 kWh was considered equal to 0.114 Euros [27] and the labor cost (k_w) equal to 32.6 Euros/hour [27]. The cost of the raw material was considered equal to the cost of its constituents (K_{mf} is the cost of carbon fibers and K_{mr} the cost of the resin). Additionally, the empirical assumption of 80% of scrap material based on Reference [4] was made for all the manufacturing processes. The energy cost for all the processes was calculated as follows, and all the CERs developed are shown in Tables 2 and 3:

$$K_i = E_i k_i \tag{3}$$

$$K_{total} = K_{labor} + K_{material} + K_{energy} + K_{recycling} \tag{4}$$

where K_i is the total energy cost, E_i is the total energy consumption and k_i is the cost of 1 kWh of electricity.

In this study, non-recurring costs are not taken into account. However, an effort is made to evaluate the equipment cost of autoclaving and RTM based on data available from a company specializing in online commerce (www.alibaba.com).

Table 2. Labor cost.

Activity	CER [11,28]
Clean the mold	$K_{cm} = k_w{}^*1{}^*2{}^*\text{PAA}$
Application of the release agent	$K_{ra} = k_w{}^*0.1{}^*\text{PAA}$
Cut prepreg (10 plies)	$K_{cu} = k_w{}^*0.25{}^*\text{PAP}$
Cut diaphragms	$K_d = 1.6k_d{}^*\text{PPA}{}^*N_D$
Lay up	$K_{layup} = k_w{}^*t_{layup}$ 8 min. for the first 2, 4 min. for the rest
Cut breather/bleeder/release film/sealing bag/peel plies	$K_{pl} = k_w{}^*0{,}25{}^*\text{PAP}$
Demolding	$K_{de} = k_w{}^*0.25{}^*\text{PAA}$
Cost of the rework	$K_{rw} = k_w{}^*0.25{}^*\text{PAP}$
Cost of the NDT inpection	$K_{ins} = k_w{}^*(1{}^*\text{PAA}{}^*\text{cmp} + 0.5)$
Cost of the dimension measurement	$K_{dim} = k_w{}^*0.5{}^*\text{cmp}{}^*\text{PAA}$
Cost of the storage	$K_{st} = k_w{}^*(0.05{}^*\text{WP} + 0.16{}^*\text{PAP})$

Table 3. Material cost.

Material	CER
Cleaning agent	$K_{cm} = k_{cm}{}^*\text{PAA}{}^*m_{cm}{}^*2$ [11]
Release agent	$K_{ra} = k_{ra}{}^*\text{PAA}/\text{Coverage}$
Breather/bleeder/release film/peel ply, sealing bag, diaphragms	$K_i = k_i{}^*L_i$
Raw material	$K_{total} = 1.8{}^*(K_{mf}{+}K_{mr}) = 1.8{}^*k_{mf}{}^*m_{mf} + 1.8{}^*k_{mr}{}^*m_{mr}$

3. Results and Discussion

3.1. Life Cycle Analysis

The results from the Life Cycle Analysis demonstrated that the production of carbon fiber is responsible for more than 80% of the total carbon footprint (Figure 3). These findings highlight the need for developing new or improve the existing recycling techniques to avoid the production of new carbon fibers. Figures 4 and 5 demonstrate that producing the canopy using a carbon fiber reinforced epoxy composite, involving RTM as the manufacturing process and mechanical recycling as the end-of-life treatment route demonstrates the lowest carbon footprint. On the contrary, the worst practice for producing the canopy is when a carbon fiber reinforced thermoplastic composite is the material of choice, and autoclaving is the manufacturing process selected because of the high embodied energy and processing temperature of PEEK as well as the higher energy demands of the autoclave equipment. The environmental performance of producing a thermoplastic canopy is improved when out of autoclave techniques (RTM and CDF) are considered as the manufacturing processes, but this amelioration is not enough to provide a lower life-cycle carbon impact.

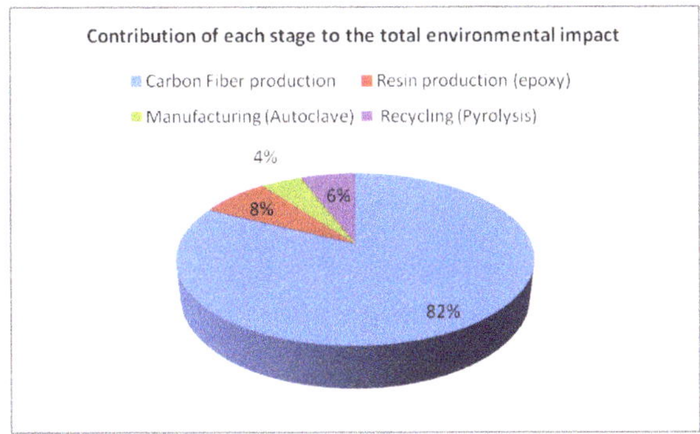

Figure 3. Contribution of each stage to the total environmental impact.

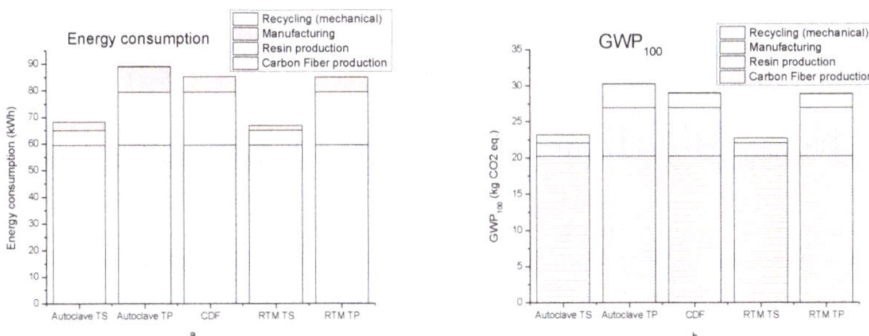

Figure 4. (**a**) Total energy consumption and (**b**) Global Warming Potential-100 for each process scenario when mechanical recycling is considered.

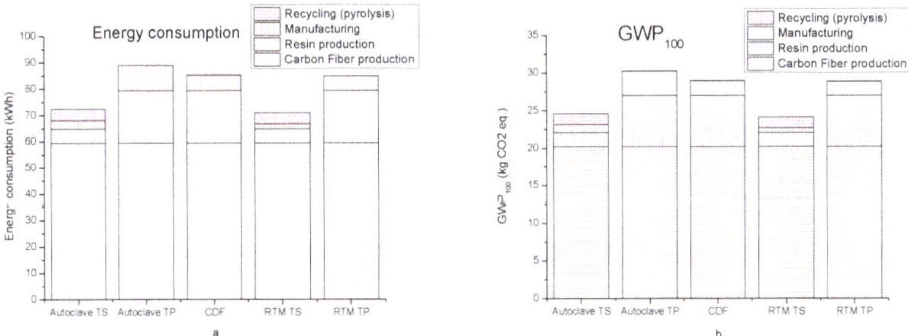

Figure 5. (**a**) Total energy consumption and (**b**) Global Warming Potential-100 for each process scenario when mechanical recycling is considered for thermoplastic and pyrolysis for thermosetting composite.

However, as already mentioned, the present study is limited to the investigation of the carbon impact without accounting for recyclability and reusability of the composite material. Therefore, it results in the paradoxical conclusion that a non-recyclable material (epoxy composite) is a more environmentally friendly alternative (lower carbon footprint) as compared to a recyclable material (PEEK composite). It should be noticed that landfill (disposing waste material by burial) is a relatively cheap disposal route and mostly used in overseas countries (e.g., USA) but is the least preferred waste management option under the European Union's Waste Framework Directive, and, therefore, opposition to it is expected to increase over the coming years. This route is already not permitted in Germany, while other EU countries are expected to follow.

It is worth noting that recycling techniques studied and implemented during the last two decades in Europe, such as mechanical processes (mainly grinding), pyrolysis, solvolysis, and other thermal processes, are not satisfactory as they either downgrade the quality of the recycled material appreciably (e.g., grinding) or are associated with enormous energy consumption (e.g., pyrolysis). To assess the efficiency of a recycling process, more holistic evaluation criteria would be needed to account for energy consumption, overall environmental impact including carbon footprint, and reusability, as well as the quality of the recycled material.

3.2. Life Cycle Costing

Figure 6 reveals that material and labor costs are responsible for 19% and 81% of the total cost, respectively. Energy and recycling costs are negligible when compared with material and labor cost. For this reason, in the following, only scenarios where mechanical recycling is considered as the recycling technique are shown. Moreover, from Figure 7, it is obvious that producing the canopy from carbon fiber reinforced epoxy composite using the RTM technique is the best option from a financial standpoint. This is the result of the lower raw material cost of the epoxy composite as compared to thermoplastic and of the absence of auxiliary materials (breather films, vacuum bags, diaphragms, etc.) for the RTM process. Furthermore, the fewer steps needed for preparing the mold decreases the labor cost of RTM by 35% and by 26% as compared to the conventional autoclave technique and CDF, respectively (Figure 8). The financial performance of the scenarios involving PEEK composite is impaired from the high raw material cost. However, it can be expected that the financial performance of the thermoplastic composites will be appreciably improved when the higher production rates offered from these materials are considered as a decision-making parameter. In addition, their high reusability potential can be taken into account.

Figure 9 demonstrates a characteristic example of the effect of the main cost drivers to the total cost of an activity. It exhibits the cost variation for storing the component as a function of two product's features, the weight and the perimeter, revealing that this activity is more sensitive to a perimeter increase than to a weight increase.

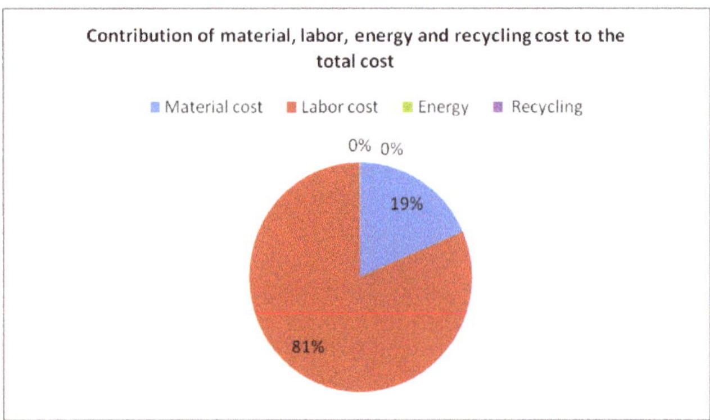

Figure 6. Contribution of the material, labor, energy and recycling cost to the total cost (autoclave-thermosetting composite).

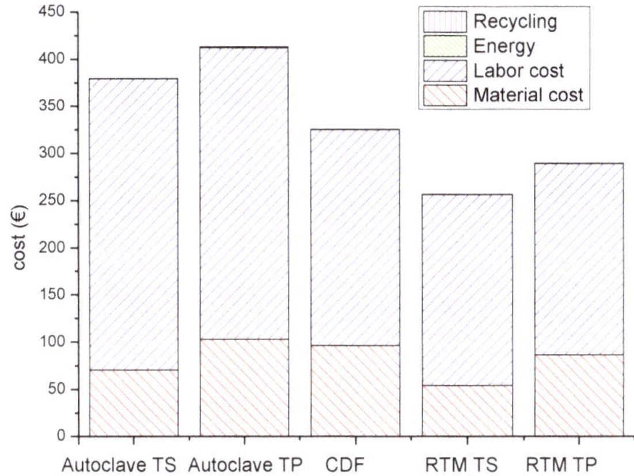

Figure 7. Total cost for the different manufacturing scenarios examined.

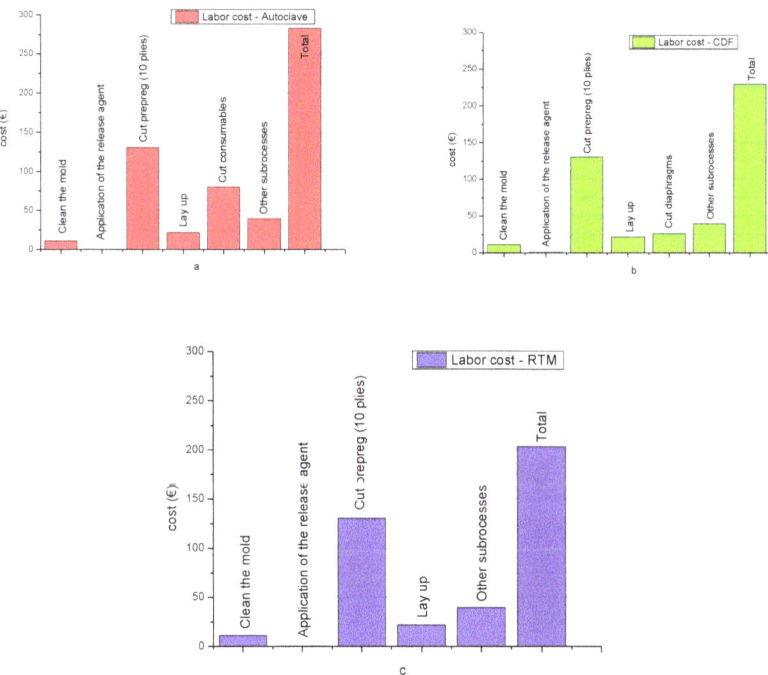

Figure 8. Labor cost breakdown structure for (**a**) Autoclave, (**b**) CDF, (**c**) RTM.

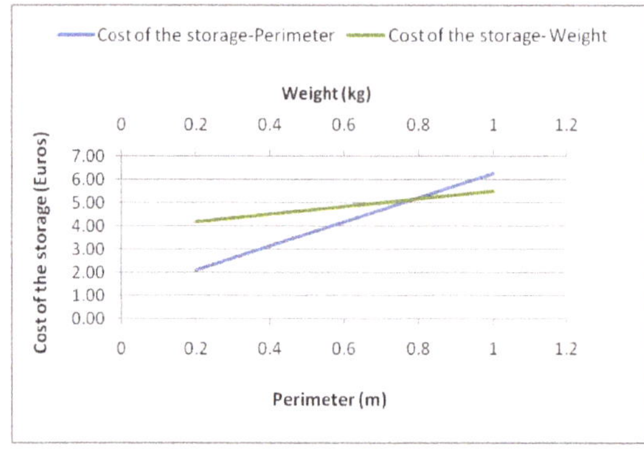

Figure 9. Cost of the storage as a function of the perimeter and of the weight.

As far as the investment cost for the manufacturing processes is concerned, this is strongly dependent on the size of the equipment, maximum working temperature and pressure as well as heating and cooling rates, injection capacity (for RTM) etc. In this frame, a cost evaluation of an autoclave and RTM (injection/mixing machine) equipment was made and is shown in Figure 10. From this figure, it is obvious that the investment cost of autoclaving is more than three times higher compared with RTM. On the other hand, increased cost of the molds used in RTM, as well as additional costs related with the design of the mold (e.g., for defining the injection line), and not taken into account in this study, impair the financial performance of RTM.

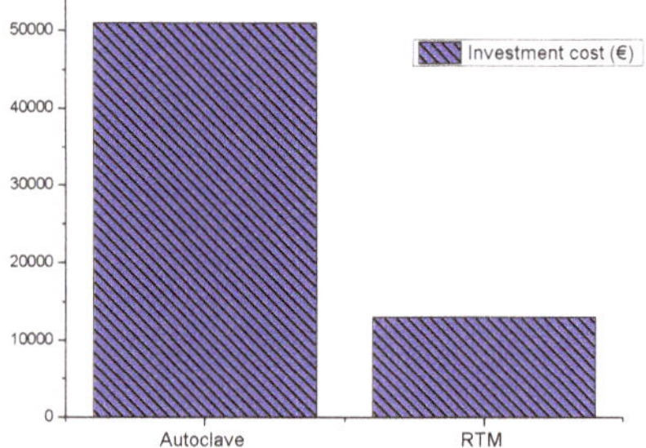

Figure 10. Investment cost of autoclaving and RTM.

4. Conclusions

In the present work a Life Cycle Analysis (LCA) combined with a Life Cycle Cost (LCC) analysis was carried out to estimate the carbon footprint and financial efficiency of different material, manufacturing, and end-of-life scenarios associated with the production of an aeronautical composite structure: the canopy of the Twin Star helicopter. The results from the environmental analysis demonstrated that the embodied energy of the raw materials and the energy intensity of each process are the key factors affecting the carbon footprint, and are the cause of the lower ecological performance of PEEK composite compared with thermosetting epoxy composite. As far as the cost analysis is concerned, the main cost drivers are the cost of the materials (raw and auxiliary materials) and the cost for preparing the mold for the process.

However, parameters not taken into account in this study, such as higher product quality referring to increased mechanical properties and lack of defects provided by autoclaving; higher productivity offered by thermoplastic composites due to reduced processing cycles; assembling potential, such as the use of welding that is expected to lead to lower assembling costs and weight reduction, as well as recyclability can affect the decision strategy at the early design stages. This also underlines the need for a holistic approach to include potential reuse and 'clear' recycling applications of the composites, as well as circular economy considerations when establishing the criteria for designing an aircraft structure, selecting the material and finally manufacturing the structure.

Finally, this analysis makes evident the need for developing a versatile concept able to both deal with intricate tasks and provide the optimal design options among different alternatives.

Author Contributions: C.V.K. was involved in the conceptualization as well as in the project administration/supervision and the review/editing of the manuscript. A.L. performed the formal analysis, the data investigation, the writing a big part of the original manuscript and the application of the methodology. S.G.P. supervised and validated the project as well as contributed to the final editing and review of the manuscript.

Funding: The present work was partially funded from EU NHYTE project.

Acknowledgments: The present work is a preliminary study performed within NHYTE project (NHYTE, 2017–2020). The NHYTE project has received funding from the European Union's Horizon 2020 research and innovation programme under grant agreement No. 723309.

Conflicts of Interest: The authors declare no conflict of interest.

References

1. Timmis, A.J.; Hodzic, A.; Koh, L.; Bonner, M.; Soutis, C.; Shafer, A.W.; Dray, L. Environmental impact assessment of aviation emission reduction through the implementation of composite materials. *Int. J. Life Cycle Assess.* **2015**, *20*, 233–243. [CrossRef]

2. Bachman, J.; Hidalgo, C.; Bricout, S. Environmental analysis of innovative sustainable composites with potential use in aviation sector—A life cycle assessment review. *Sci. China Technol. Sci.* **2017**, *60*, 1301–1317. [CrossRef]

3. Duflou, J.R.; Deng, Y.; Van Acker, K.; Dewulf, W. Do fiber-reinforced polymer composites provide environmentally benign alternatives? A life-cycle-assessment-based study. *MRS Bull.* **2012**, *37*, 374–382. [CrossRef]

4. Song, Y.S.; Youn, J.R.; Gutowski, T.G. Life cycle energy analysis of fiber-reinforced composites. *Compos. Part A* **2009**, *40*, 1257–1265. [CrossRef]

5. Michaud, V. Les Matériaux Composites, Moteurs de la Mobilité Proper. Presented at the Swiss Mobility Days, Martigny, Switzerland, 7 April 2016.

6. Deng, Y. Life Cycle Assessment of Biobased Fibre-Reinforced Polymer Composites. Ph.D. Thesis, Katholieke Universiteit Leuven, Leuven, Belgium, 2014.

7. Das, S. Life cycle assessment of carbon fiber-reinforced polymer composites. *Int. J. Life Cycle Assess.* **2011**, *16*, 268–282. [CrossRef]

8. Hueber, C.; Horejsi, K.; Schledjewski, R. Review of cost estimation: Methods and models for aerospace composite manufacturing. *Adv. Manuf. Poly. Compos. Sci.* **2016**, *2*, 1–13. [CrossRef]

9. Bader, M.G. Selection of composite materials and manufacturing routes for cost effective performance. *Compos. Part A* **2002**, *33*, 913–934. [CrossRef]

10. Barlow, D.; Howe, C.; Clayton, G.; Brouwer, S. Preliminary study on cost optimization of aircraft composite structures applicable to liquid moulding technologies. *Compos. Struct.* **2002**, *57*, 53–57. [CrossRef]

11. Pantelakis, S.G.; Katsiropoulos, C.V.; Labeas, G.N.; Sibois, H. A concept to optimize quality and cost in thermoplastic composite components applied to the production of helicopter canopies. *Compos. Part A* **2009**, *40*, 595–606. [CrossRef]

12. Soares, S.R.; Finotti, A.R.; da Silva, V.P.; Alvarenga, R.A. Applications of life cycle assessment and cost analysis in health care waste management. *Waste Manag.* **2013**, *33*, 175–183. [CrossRef] [PubMed]

13. Simões Carla, L.; Costa Pinto Lígia, M.; Bernardo, C.A. Environmental and economic assessment of a road safety product made with virgin and recycled HDPE: A comparative study. *J. Environ. Manag.* **2013**, *114*, 209–215. [CrossRef] [PubMed]

14. Miro, R.; Antti, S.; Jukka, H.; Seppo, J. Combining life cycle costing and life cycle assessment for an analysis of a new residential district energy system design. *Energy* **2013**, *63*, 168–179.

15. Antonella, P.; Fabio, D.; Elio, J.; Claudio, A.; Mariagiovanna, M.; Lavadera, L.A. Life cycle assessment (LCA) and life cycle cost (LCC) analysis model for a stand-alone hybrid renewable energy system. *Renew. Energy* **2016**, *95*, 337–355.

16. Norris, G.A. Integrating Life Cycle Cost Analysis and LCA. *Int. J. LCA* **2001**, *6*, 118–120.

17. Aysun, Ö.; Zerrin, G.; Gülden, T.; Levent, K.; Melike, M.; Müfide, B.; Alpagut, K. Life Cycle Assessment and Life Cycle Cost Analysis of Magnesia Spinel Brick Production. *Sustainability* **2016**, *8*, 662.

18. Koronis, G.; Silva, A. Green Composites Reinforced with Plant-Based Fabrics: Cost and Eco-Impact Assessment. *J. Compos. Sci.* **2018**, *2*, 8. [CrossRef]

19. Xiang, L.; Ruibin, B.; Jon, M. Environmental and financial performance of mechanical recycling of carbon fibre reinforced polymers and comparison with conventional disposal routes. *J. Clean. Prod.* **2016**, *127*, 451–460.

20. Robert, A.W.; Jʋrʈme, P.; Veronique, M.; Christian, L.; Jan-Anders, E.M. Assessing the life cycle costs and environmental performance of lightweight materials in automobile applications. *Compos. Part A* **2011**, *42*, 1694–1709.

21. ISO 14040. *Environmental Management- Life Cycle Assessment-Principles*; ISO: Geneva, Switzerland, 2006.

22. Energetics Incorporated. *Bandwidth Study on Energy Use and Potential Energy Saving Opportunities in the Manufacturing of Lightweight Materials: Carbon Fiber Reinforced Polymer Composites*; U.S. Department of Energy: Washington, DC, USA, 2016.

23. Suzuki, T.; Takahashi, J. Prediction of energy intensity of carbon fiber reinforced plastics for mass-produced passenger car. In Proceedings of the 9th International SAMPE Symposium, Tokyo, Japan, 29 November–2 December 2005.

24. Howarth, J.; Mareddy, S.S.R.; Mativenga, P.T. Energy intensity and environmental analysis of mechanical recycling of carbon fibre composite. *J. Clean. Prod.* **2014**, *81*, 46–50. [CrossRef]

25. Job, S.; Leeke, G.; Mativenga, P.T.; Oliveux, G.; Pickering, S.; Shuaib, N.A. *Composites Recycling: Where Are We Now*; Composites UK Ltd.: Berkhamsted, UK, 2016.

26. Oiveux, G.; Dandy, L.O.; Leeke, G.A. Current of recycling of fibre reinforced polymers: Review of technologies, reuse and resulting properties. *Prog. Mater. Sci.* **2015**, *72*, 61–99. [CrossRef]

27. Electricity Map: Live CO_2 Emissions of Electricity Consumption. Available online: www.electricitymap.org (accessed on 1 September 2018).

28. Tong, R.; Hoa, S.V.; Chen, M. Cost Analysis on L-shape Composite Component Manufacturing. In Proceedings of the 18th International Conference on Composite Materials, Jeju, Korea, 21–26 August 2011.

Article

Manufacturing Process of High Performance–Low Cost Composite Structures for Light Sport Aircrafts

Mauricio Torres [1],* , Saúl Piedra [1], Saúl Ledesma [2], Carlos A. Escalante-Velázquez [1] and Giovanni Angelucci [3]

[1] National Council of Science and Technology (CONACYT)—National Center for Aeronautics Technologies—Center for Engineering and Industrial Development (CENTA-CIDESI), Colón, Querétaro 76269, Mexico; saul.piedra@cidesi.edu.mx (S.P.); carlos.escalante@cidesi.edu.mx (C.A.E.-V.)

[2] National Centre for Aeronautics Technologies (CENTA), Colón, Querétaro 76270, Mexico; saul.ledesma@cidesi.edu.mx

[3] Horizontec S. A. de C. V., 01700 Ciudad de Mexico, Mexico; angeluccigio@gmail.com

* Correspondence: maothor@gmail.com or mauricio.torres@cidesi.edu.mx; Tel.: +52-442-211-9800 (ext. 5333)

Received: 13 December 2018; Accepted: 24 January 2019; Published: 1 February 2019

Abstract: This work describes the technological and scientific efforts on designing, manufacturing and testing validation for high performance-low cost composite structures for Light Sport Aircrafts (LSA). A Mexican initiative to conceive, manufacture and assembly a Light Sport Aircraft has been developed by using Computational Fluid Dynamics (CFD), Finite Element Analysis (FEA) and Liquid Composite Manufacturing (LCM). These consolidated techniques are used to characterize novel approaches to manufacturing and assembly carbon-fiber based structural components. As large structures are manufactured via Vacuum Assisted Resin Infusion (VARI), impregnation strategies are studied to minimize inner flaws and also to improve the manufacturing time and surface quality of each component. The first case of study, to validate this methodology, involves non-structural components such as the cowling. Control surfaces (ailerons, rudder, elevator and flaps) have been manufactured, each of them having common issues but also unique challenges. As an example, a second case of study, the aileron main beam is analyzed. Furthermore, test portfolio will be developed with the goal to perform 1-to-1 scale mechanical tests for validation in compliance with ASTM standards.

Keywords: light sport aircraft; composites structures; vacuum assisted resin infusion; building-block approach

---

1. Introduction

In the frame to develop small aircraft technology in Mexico, the company Horizontec in a joint venture with CENTA propose the design, manufacturing and test-in-fly of Light Sport Aircrafts (LSA). Halcon 1 airplane (Falcon 1) was developed using aeronautic wood and glass fiber reinforced plastics (GFRP). Nowadays, it is a full certified aircraft, under mandatory circular for Mexican light and experimental aircrafts [1].

The Halcon 2 project (Falcon 2) aims at designing and manufacturing a LSA made out of carbon fiber reinforced plastics (CFRP). The goal is the certification of the aircraft in accordance with the mandatory circular for Mexican light and experimental aircrafts [1] and the ASTM Standard F2245 [2].

Like any aircraft structure, the Halcon 2 project is designed and manufactured under the "Building-Block Approach," known as the Pyramid of Tests [3,4]. This design methodology deals with several questions on the comprehension of new structural concepts that are not still fairly answered. As an example, the major causes of in-service damage of composite structures are manufacturing process miscues and low velocity impacts [5,6]. These flaws can promote visible induced damage (BVID), which is the prime source of components debonding, delamination, ply fracture and fiber cracking in composite structures. These failure mechanisms are much more complex than those of conventional metallic materials, leading to new problems for maintenance tasks and repair procedures. Even when numerous analytical, experimental and numerical efforts have been done with the aim to forecast stress, strain and failure mechanics for composite structures, the majority of these studies have been carried out at laboratory scale, where analytic accuracy is still reasonable although budget and infrastructure restrictions [7–10].

The pyramid of tests deals with five levels of mechanical tests [1]. In the first step, simple coupons are tested in order to obtain basic properties for new and/or modified materials. The second level is conceived to design structural basic elements such as plates, shells, beams and stiffeners. The third level is defined by structural details such as plates with drop-offs, stiffened panels or wing-boxes. The fourth level is dedicated to analyze complete components like a wing, elevator or fuselage. Finally, the fifth level considers testing of the whole aircraft.

The higher the level is accomplished, the bigger the operational and financial risks are taken. Normally, the analysis of singularity details on composite structures, such as ply drop-offs, bolted/fastened/glued joints or edge effects, takes place at the third and fourth levels. At those stages, if a concept or material problem is detected, it is difficult and costly to change the complete design [4].

In the case of the Halcon 2, the key challenge is to manufacture all main components of the aircraft by Vacuum Assisted Resin Infusion (VARI). Other efforts for manufacturing aircraft components in "single step" have become reliable in recent years [11–14]. Meredith et al. [10] have studied the performance versus cost analysis of carbon fiber reinforced plastics (CRFP), for a specific requirement such as specific energy absorption (SEA). By comparing autoclave versus conventional oven cures, they were found to have identical SEA, providing a cost benefit for massive production. Another example was reported by Verma et al. [11] by analyzing the challenges in a cocured wing test box by vacuum enhanced resin infusion technology (VERITy). They focused on four major technological challenges: 1) selection of materials and optimization of process parameters; 2) design and development of tools; 3) layup of dry layers and assembly preforms; and 4) design of infusion strategy. They concluded the importance of following the Building-Block Approach to get a proper infusion, as well as to improve the numerical tools to evaluate the flow front and to describe the uncertainties associated with complex components. Finally, on the same order of ideas, Komarov et al. [12] evaluated shape distortion during all stages of the vacuum infusion production of composite aerospace structures. They cited that high quality numerical tools (FEA and CFD) for molding products is important for achieving the required accuracy of manufactured products.

Within this framework, the Building-Block Approach is applied to the H2 aircraft as follows. For Level 1, coupon level, a series of test for each material used on the LSA must be done. An extensive materials characterization was performed for obtaining the elastic constants for the composite materials. For Level 2, elements level, performance of sandwich structures is required, since most key elements of the fuselage and wings will use this configuration. These properties have been used in the FEA as input, for taking decisions on the configuration of the different Halcon 2 components and to describe the baseline structural performance. For levels 3 and 4—components and subsystems level—the design of mold and components relies on CFD approaches to calculating the flow front behavior and the resin infusion strategy [13]. Here, surface controls (ailerons, rudder, elevator and flaps) have been manufactured, as well as non-structural components (cowling). Efforts are now devoted to the manufacturing and assembly of the wing and fuselage. Moreover, at Level 5, aircraft level, test frames are now being developed with the goal to perform 1-to-1 scale mechanical tests on the assembled aircraft.

In the present paper, two case studies: 1) cowling and 2) aileron main beam are described to show the engineering processes and the technological challenges encountered in manufacturing high performance-low cost composite structures for Light Sport Aircrafts.

2. Materials and Methods

The main goal of this work is to describe the use of computational and engineering tools for confronting the challenges of manufacturing low cost composite components for a light sport aircraft. It is well known that any aeronautical development must comply with a series of specific stages in order to get an airworthiness certification. These stages are prescribed by the so-called Building-Block Approach [3,4]. After product design, manufacturing processes must be implemented and also validated through experimental tests. Here, two different case studies are shown about the design and verification of the manufacturing processes which is being developed by Horizontec Company. Firstly, it is shown how Computational Fluid Dynamics simulations can be performed to simulate the impregnation process of an aircraft component. The development of simple but consistent CFD models to simulate the VARI process are discussed in the next sections. Also, experimental observations were carried out in order to calculate relevant input parameters for the CFD simulations to increase the reliability of the results. Once the manufacturing process is developed and the component is fabricated, the performance must be validated through mechanical testing. For composite materials, this verification stage is more complex than for traditional materials. Although in the preliminary design and detailed engineering stages, the configuration of materials for each component is determined by computer simulations; the effect of the manufacturing process on the mechanical performance must be studied by means of mechanical tests of each component. The second case study presented in this work shows how it is possible to compare the results from a simple mechanical testing with a computational simulation and use the results of such comparison to verify if the manufacturing process has any effect on the mechanical performance of a structural element of the Light Sport Aircraft.

2.1. Materials

Materials employed for the manufacturing of H2 aircraft are listed in Tables 1–3. For the cowling, AS4 Spreadtow carbon fibers by Hexcel® (Seguin, TX, USA) are used. For the aileron main beam, IM7 carbon fibers (unidirectional (UD) and biaxial fabric) by Hexcel® and PVC foam core by 3D Core® (Herford, Germany) are used. In both cases, EPOLAM 2019 epoxy resin by Axson® (San Luis Potosi, Mexico) is employed.

Table 1. Reinforcement materials employed in H2 cowling and aileron main beam.

Material [1]	Density (kg/m^3)	Tensile Modulus (GPa)	Tensile Strength (MPa)	Filament Diameter (micron)	Carbon Content (%)
AS4	1.79	231	4400	7.1	94
IM7	1.78	276	5600	5.2	95

[1] All values from material's data sheet [15,16].

Table 2. Epoxy resin used in H2 components.

Property	Value [1]
Density cured (kg/m^3)	1.12
Mixing ratio by weight	100/35
T_g (°C)	110
Curing (h @ °C)	16 @ 25 °C
Viscosity mixed at 25 °C (mPa·s)	250
Pot life (500 g) at 25 °C (min)	95

[1] All values from material's data sheet [17].

Table 3. PVC foam core for H2 components.

Property	Value [1]
Density (kg/m^3)	60
Thickness (mm)	3–15
Maximum operational temperature (°C)	70
Young's Modulus (MPa)	70
Compression Strength (MPa)	4.5
Application for vacuum infusion	Very good

[1] All values from material's data sheet [18].

2.2. Manufacturing of CFRP and Sandwich Coupons

CFRP and sandwich plates are manufactured by Vacuum Assisted Resin Infusion (VARI) (Horizontec-CENTA, Queretaro, Mexico) (Figure 1a). For characterizing AS4 and IM7 CFRP, UD and $[\pm 45]_4$ coupons are employed. For characterizing CFRP-PVC sandwiches, the $[0_2/\text{PVC}/0_2]$ layout is used.

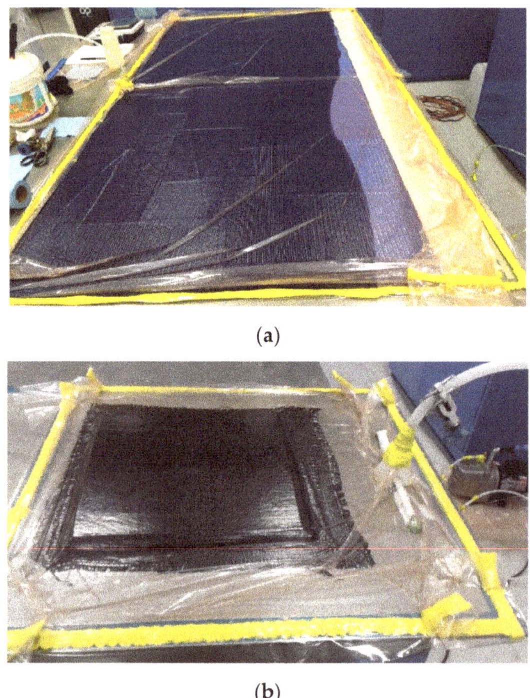

(a)

(b)

Figure 1. Vacuum Assisted Resin Infusion for (**a**) Carbon fiber reinforced plastic (CFRP) coupons and (**b**) example of CFRP-PVC foam core sandwich.

The vacuum pressure is fixed at 0.25 bar for 2 h, till gelation of the resin is evident. The polymerization reaction takes place for 16 h at 25 °C. For the CFRP, post-curing of 2 h at 60 °C, then 2 h at 80 °C and finally 4 h at 100 °C, following the recommended recipe by Axson® (Figure 1b) [17]. Sandwich structures were post-cured for 2 h at 60 °C [17].

2.3. Mechanical Characterization of CFRP and Sandwich Coupons

For the CFRP, mechanical characterization was carried out following ASTM D3039 and ASTM D3518 standards [19,20]. Nominal dimensions for the carbon-epoxy coupons were 250 × 25 × 1 mm with a testing zone restricted to 150 mm. All specimens were instrumented with strain gauges at the surface. For tabs, 60 grain gridding paper was used. For the tensile tests, MTS Insight 100kN Universal Testing Machine (CENTA, Queretaro, Mexico) was employed. The crosshead rate was set to 1mm/min until coupon failure is reached (Figure 2a).

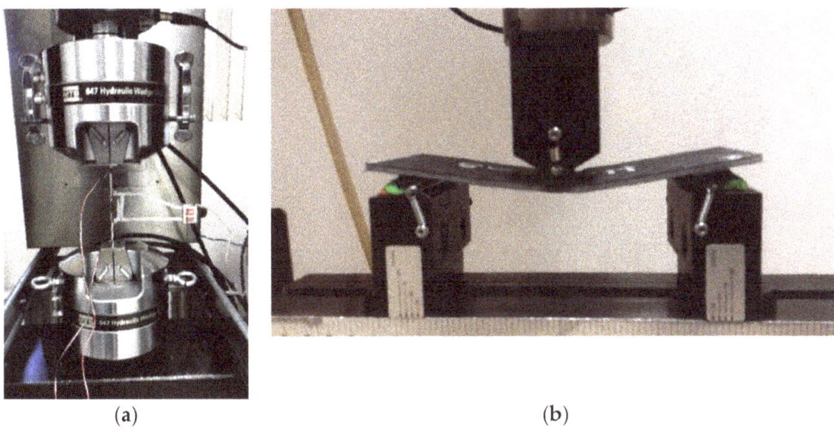

(a)   (b)

Figure 2. Mechanical characterization for (**a**) CFRP coupons, (**b**) CFRP-PVC foam sandwich coupons.

For the CFRP-PVC sandwich, flexural tests according to ASTM C393 standard [21] were performed. Nominal dimensions for the sandwich were 200 × 75 mm. The average thickness for the sandwich samples was 3.5 mm. The span length was fixed in 150 mm. For the flexural tests, an Instron 8872 Servohydraulic Testing Machine (CENTA, Queretaro, Mexico) was employed. The crosshead rate was set to 6mm/min until the coupon failure is reached (Figure 2b).

2.4. CFD Modeling for the Impregnation Process of the Aircraft Upper Cowling

The quality of any composite component manufactured by VARI process depends strongly on the impregnation stage. During this stage, the resin flows through the preforms of carbon fiber and the pores must be fully filled by the resin in order to avoid defects in the final product. In this case study, we present the implementation of CFD simulations to compute the global behavior of the resin flow, estimate the filling time of the component and verify if the resin inlet and outlet gates are appropriate to carry out the VARI process. For this purpose, the analysis of the resin flow through the upper cowling of a light sport aircraft is developed. In recent years, CFD models to predict the resin flow through different technical fabrics have been studied in order to use them as a manufacturing process design tool, for instance [22–24]. There are multiscale models that can take into account different effects during the resin impregnation stage, however, as an engineering tool, the CFD models must be as simple as possible in order to be used for testing different configurations of the VARI process and selecting the impregnation strategy for manufacturing the component.

The CFD model for the resin flow through the cowling fibers preform is based on the solution of the porous media equations in ANSYS Fluent. The Volume-of-Fluid (VOF) technique was implemented in order to solve the equations for a two-phase flow (air-resin) including the interface and to track the advance of the resin front as a function of time [25]. An unstructured tetrahedral mesh was built to discretize the conservation equations. The time step was adjusted throughout the simulation in order to obtain a Courant number less than 1 during the entire simulation. The boundary conditions implemented in the model were based on the vacuum pressure that the company has fixed for its VARI processes. The geometry of the aircraft upper cowling is shown in Figure 3a. Three different configurations for the inlet and outlet gates were simulated, C-1, C-2 and C-3. The inlet gate for each case is plotted in red color in Figure 3a, the resin outlet was simulated in the lateral edges of the cowling. The laminate code for manufacturing the upper cowling is shown in Figure 3b.

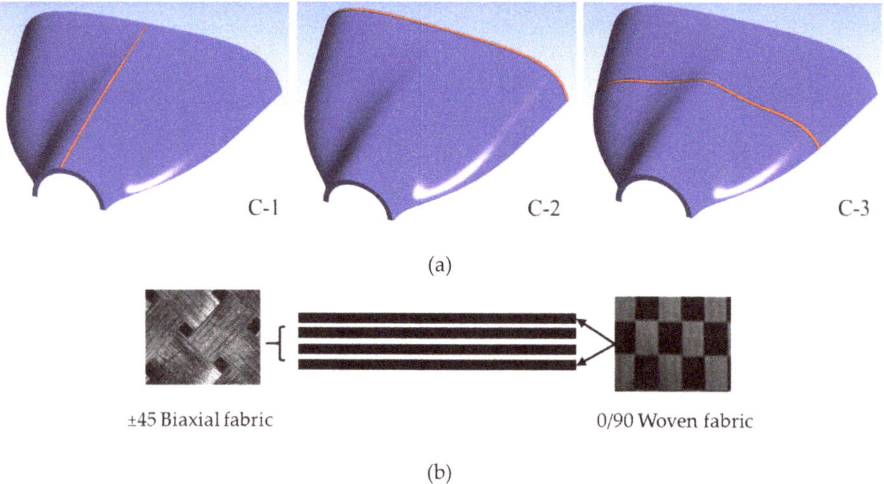

Figure 3. (**a**) Geometry of the aircraft upper cowling; (**b**) Laminate code.

The critical part for simulations of these kinds of flows is the calculation of the porosity and permeability of the preform laminate configuration. As was commented, since in the CFD model the preforms are solved as a porous medium, the porosity and permeability tensor must be incorporated into the numerical model. The porosity of the medium can be found through fiber volume fraction experiments for individual layer specimens of each kind of fiber that comprises the laminate. Once the porosity of each ply is calculated, the effective porosity of the laminate is computed as an average of the porosity of the individual layers weighted by their thickness.

The permeability tensor is very difficult to compute; almost all the simulations reported in the literature are solved using the permeability found through theoretical models or experimental observations. The theoretical models to compute the permeability using the geometrical characteristics of the fiber have been developed to provide this information to numerical simulations; however, the permeability depends in many parameters that involves the geometrical features but also the fluid that moves through the porous medium. As a result, the models reported do not predict accurately the permeability of a laminate. In this work, experimental observations were carried out to estimate the permeability of each individual ply of the laminate. Once the individual permeability was calculated, the effect permeability of the preform was computed by averaging the individual values. The experiments used to calculate the effective permeability of the porous medium are based on a vacuum assisted resin transfer (VARTM) setup as shown in Figure 4 [26]. The rigid part of the mold is made from a rectangular $1.2 \times 1.5 \times 0.01$ m^3 glass plate on which preforms up to 0.2×1 m^2 in area were laid up. Given the high degree of repeatability that can be achieved, a linear channel flow experiment driven by a constant pressure gradient was chosen for the measurement of permeability [23]. Moreover, the preform width was chosen to be about 16 times the inlet hose diameter. In this manner, good practices for permeability measurement are observed [27].

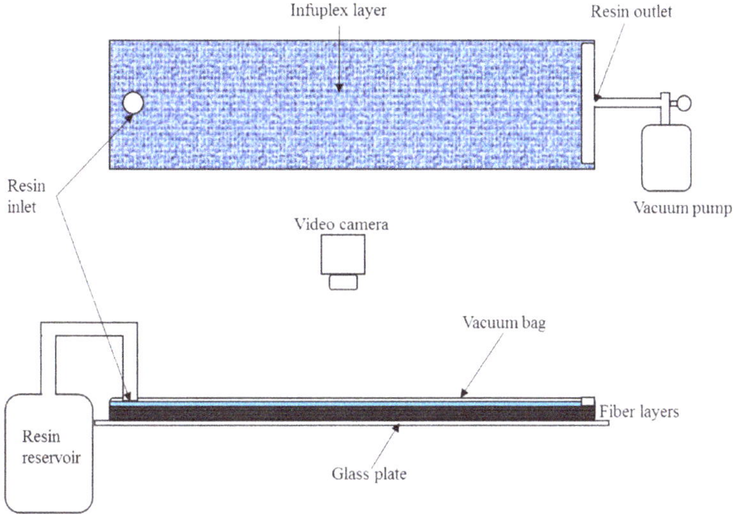

Figure 4. Experimental set up for manufacturing the laminates.

The mold and preform preparation proceeded as follows. First, the glass plate was treated with mold cleaning (acetone) and release agents (demolding wax). Next, carbon fiber fabrics were cut to 0.1×0.9 m^2 and laid up on the glass. An infuplex layer (high permeability layer, HPL) was placed on top of the carbon fiber fabrics in order to promote a uniform resin flow along the laminate. Afterwards, a point injection source and a linear sink were placed on the glass plate in such a manner that the resin would flow along the longest preform dimension. After the vacuum bag was placed and sealed, the linear sink and the injection source were connected to the vacuum pump and resin reservoir, respectively. Finally, a digital video camera was mounted above the glass plate to monitor the flow front progression with time. The vacuum pump maintained a pressure of 3.07 kPa throughout the resin transfer process. The results obtained are shown in the following section.

2.5. Experimental Test and FEM Simulation for Aileron Main Beam

For structural elements of the aircraft components, it is mandatory to verify the manufacturing process and then the quality of each element. Following the ASTM standards applicable for this kind of aircraft, a road map of the experimental tests for the elements of the structural components was developed [2]. The first structural component manufactured was the aileron and as it is well known, the critical element of this component is the main beam. The beam was designed with a rectangular cross-section using two different fiber fabrics and a foam core. In Figure 5, the materials configuration and the cross-section of the beam are shown. The layout configuration for the aileron beam consists in a central PVC foam core, three unidirectional layers in each beam flange (top and bottom) and two [±45] fabrics around them.

The performance of the aileron main beam was evaluated in cantilever conditions and the results were compared with the ones predicted by a Finite Element Method (FEM) simulation (CENTA, Queretaro, Mexico). The experimental test consisted of a fixed cantilever beam and an applied localized load, P, which was increased until the beam failed. The FEM simulation was implemented using ANSYS ACP utilizing the mechanical properties obtained from the level 1 tests. A total mesh of 8640 elements was constructed for the fiber-core assembly. Shell type elements were used for fiber layers and Solid type elements for the PVC foam core.

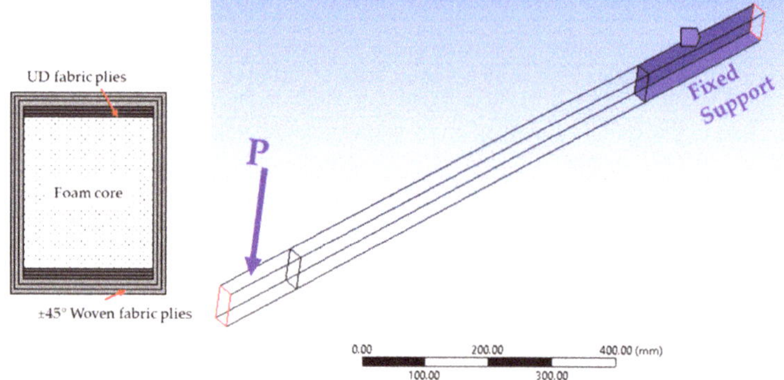

Figure 5. Materials configuration sketch and computational model for the aileron main beam.

3. Results

3.1. Mechanical Testing for Materials Characterization

Mechanical characterization for both CFRP and CFRP-PVC sandwiches will provide the elastic properties to evaluate the performance of structural components for the H2 aircraft. The elastic properties of the CFRP and CFRP-PVC foam core sandwich are given in Tables 4–7.

Table 4. Mechanical properties of AS4 Spreadtow-EPOLAM 2019 CFRP used in H2 cowling.

Elastic Property	Mean Value [1]
Young's Modulus, E_{xx} (GPa)	63.9
Poisson ratio, v_{xy}	0.033
Ultimate tensile strength, $\sigma_{x\ max}$ (MPa)	1233.3
Shear Modulus, G_{xy} (GPa)	4.4
Maximum in-plane shear stress, $\tau_{xy\ max}$ (MPa)	42.4

[1] All values from experimental data following ASTM D3039 [19] and ASTM D3518 [20].

Table 5. Mechanical properties of IM7/EPOLAM 2019 CFRP used in aileron main beam.

Elastic Property	Mean Value [1]
Longitudinal Young's Modulus, E_{xx} (GPa)	128
Transverse Young's Modulus, E_{yy} (GPa)	7.3
Poisson ratio, v_{xy}	0.3
Ultimate tensile strength, $\sigma_{x\ max}$ (MPa)	1500
Shear Modulus, G_{xy} (GPa)	5.1
Maximum in-plane shear stress, $\tau_{xy\ max}$ (MPa)	50

[1] All values from experimental data following ASTM D3039 [19] and ASTM D3518 [20].

Table 6. Mechanical properties of IM7 Biax/EPOLAM 2019 CFRP used in aileron main beam.

Elastic Property	Mean Value [1]
Young's Modulus, E_{xx} (GPa)	87
Poisson ratio, v_{xy}	0.025
Ultimate tensile strength, $\sigma_{x\ max}$ (MPa)	1100
Shear Modulus, G_{xy} (GPa)	3.5
Maximum in-plane shear stress, $\tau_{xy\ max}$ (MPa)	100

[1] All values from experimental data following ASTM D3039 [19] and ASTM D3518 [20].

Table 7. Mechanical properties of AS4 Spreadtow-EPOLAM 2019-PVC sandwich.

Elastic Property	Mean Value [1]
Core Shear Ultimate Strength, Fs_{ult} (MPa)	0.8
Facing Stress, σ (MPa)	480
Maximum force prior to failure, P_{max} (N)	420

[1] All values from experimental data following ASTM C393 [21].

For all CFRP coupons, the coordinate systems referred is the following: x: direction of the tensile load, y: transverse direction of the tensile load and z: through-the-thickness direction of the laminate. Young's Modulus was calculated by using the chord modulus equation in the linear portion of the stress-strain curve, between 1000 $\mu\varepsilon$ and 3000 $\mu\varepsilon$ [19]. Shear Modulus was calculated using the chord modulus equation in the linear portion of the stress-strain curve, below 4000 $\mu\varepsilon$ [20].

For the sandwich coupons, the coordinate systems referred is the following: x: direction of the sandwich span, y: direction of the sandwich width and z: through-the-thickness direction of the sandwich. Core shear ultimate strength and facing stress were calculated by using the proper equations of the ASTM C393 standard [21].

Figure 6 shows the IM7/EPOLAM 2019 CFRP stress-strain curves from tensile test following ASTM D3039 standard. As expected, on one side, stress-strain curves for UD coupons show a linear elastic behavior at the beginning of the plots, below 6000 $\mu\varepsilon$, where mechanical properties are calculated (Figure 6a). Nonetheless, above 10,000 $\mu\varepsilon$, the material exhibits a notable elastoplastic behavior, prior to failure. On the other side, stress-strain curves for $[\pm45]_{2s}$ coupons, above 3000 $\mu\varepsilon$, exhibit a more evident elastoplastic behavior (Figure 6b).

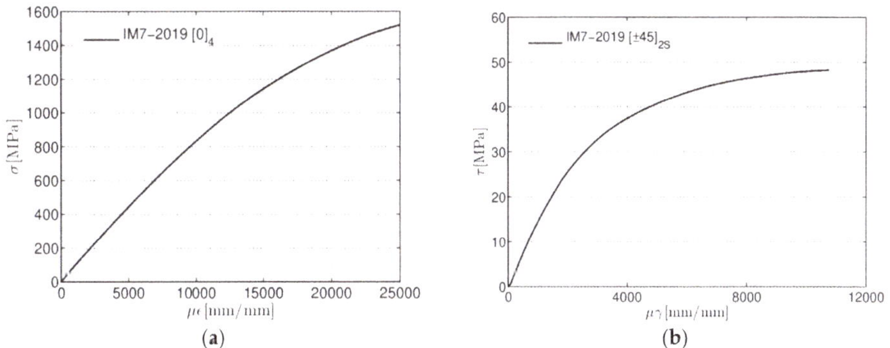

(a)   (b)

Figure 6. Tensile stress-strain curves for IM7-2019 composite (a) unidirectional (UD) [0]4 and (b) $[\pm45]_{2S}$.

On one hand, the failure mode for UD and biaxial coupons is on sudden fracture type LGV (Lateral, Gage, Various, Figure 7a). The fracture was explosive, where most of the CFRP fragments fell down inside the protection chamber. On the other hand, failure mode for [±45] coupons is warp-weft sliding, type SGT (Splitting, Gage, Top, Figure 7b). Both failure modes are according to typical codes of ASTM D3039. The failure mode for the CFRP sandwich with PVC core was in the compressed facing type FGT (Facing failure, Gage, Top facing, Figure 7c), with barely visible core crushing.

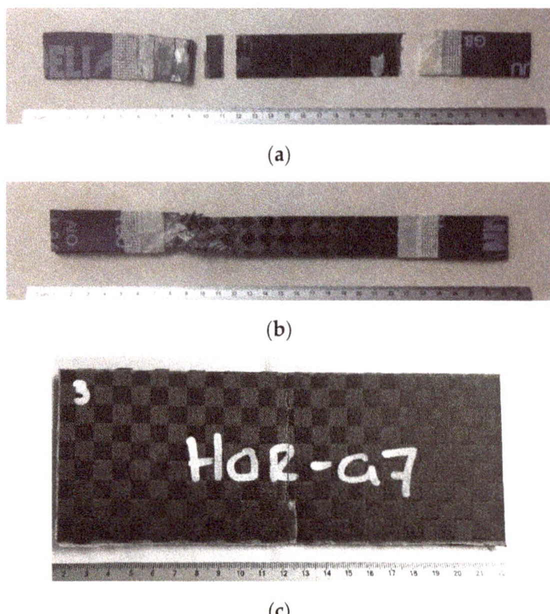

(a)

(b)

(c)

Figure 7. Failure modes for AS4-2019 composite (**a**) [0]4 and (**b**) [±45]₂S and (**c**) CFRP-PVC sandwich.

3.2. Cowling Impregnation Process

Before the CFD implementation, the permeability characterization was performed for the laminate configuration selected for manufacturing the upper cowling of the aircraft. Using the experimental setup described in Section 2.1, analysis of resin flow through each type of fiber was carried out. An image processing algorithm was used to compute the advance of the front as a function of time. Knowing the pressure gradient and the viscosity of the resin, the permeability of the fiber was calculated through the integration of the Darcy's Law [28]. The same procedure is being implemented for all the components of the Light Sport Aircraft, so that the experiments to characterize the resin flow through the fiber selected to manufacture each component of the aircraft were performed through this kind of observations. In Figure 8, one can see the resin front advance as function of time for the ±45 biaxial fiber. At the beginning of the impregnation process, a radial flow is developed through the carbon fiber plies according to the point source configuration in the injection gate. However, the lateral boundary constraints force the radial flow to evolve into a linear flow in about 90 seconds. From that point and until the fiber laminate is totally impregnated by resin, the flow advance is linear and a Darcy Law in one direction can be applied to compute the permeability.

Once the permeability of each ply of the laminate was determined, the average effective permeability and porosity were calculated for being included as an input parameter of the CFD model. The effective permeability resulting from the experiments for the cowling laminate was $k_{eff} = 3.8641 \times 10^{-12}$. The details with respect to the permeability calculation from the experiments for each ply of the laminate can be found in Reference [26]. Fiber volume fraction tests were performed for each ply of the laminate in order to compute the porosity. The effective porosity of the laminate was calculated as the average porosity weighted with the thickness of each ply, resulting equal to $\phi_{eff} = 0.479$. The filling process simulation were carried out for the three configurations C-1, C-2 and C-3, presented in Figure 3. In order to analyze quantitatively the filling process of the component, the integral of the resin volume fraction as function of time was computed during the simulations and is plotted in Figure 9 for the three configurations. The curves of the resin volume fraction can give relevant information about the evolution of the impregnation process. As can be observed, with the C-1 configuration the cowling fills faster than with the others. The filling process with C-2 configuration lasts much more time that the C-1 and C-2 and after 1750 s in the simulation, the cowling was not completely filled of resin. On the other hand, the resin volume fraction as a function of time behaves in a very similar way to the beginning of the process for C-1 and C-3 configurations, however, in the final part of the process, the resin impregnation of the preform slows down in C-3 with regarding C-1. From this information, the C-1 configuration can be considered the best choice in order to minimize the filling time. For this configuration, the resin flows and fills around 40% of the component very quickly until the flow considerably slows down due to the pressure balance inside the porous medium.

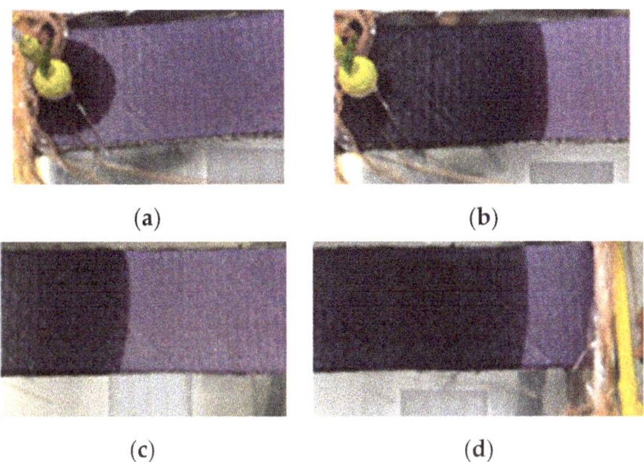

Figure 8. Sequence of representative images of flow fronts in the laminate manufacturing under vacuum assisted resin transfer (VARTM) process: (**a**) t = 10 s; (**b**) t = 100 s; (**c**) t = 1080 s and (**d**) t = 2055 s.

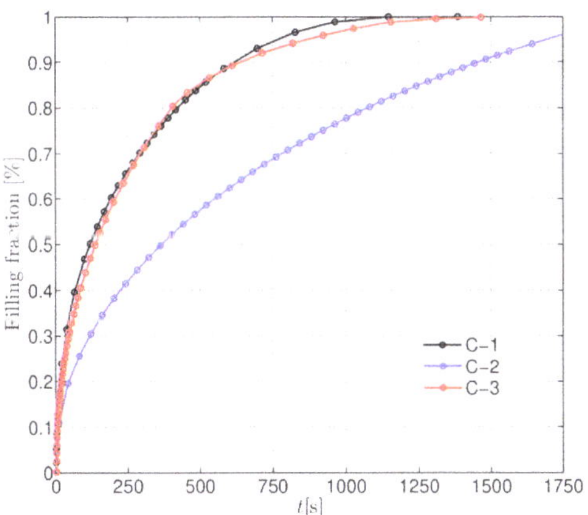

Figure 9. Filling fraction as function of time for the different inlet gate configurations gate.

In Figure 10, the filling process of the upper cowling is shown, through the contour map of the resin volume fraction for C-1 configuration. As it can be appreciated, the resin flows very fast at the beginning of the process and the velocity of the resin front decreases as the preform is filled. The results of the simulation shows that using this configuration, the component fills uniformly and symmetrically with respect to its longitudinal axis. The total filling time for the component was about 1180 s. In Figure 10d, the VARI process implemented in the laboratory for the cowling using the C-1 configuration is shown, it is important to highlight that the component successfully manufactured with good surface quality results.

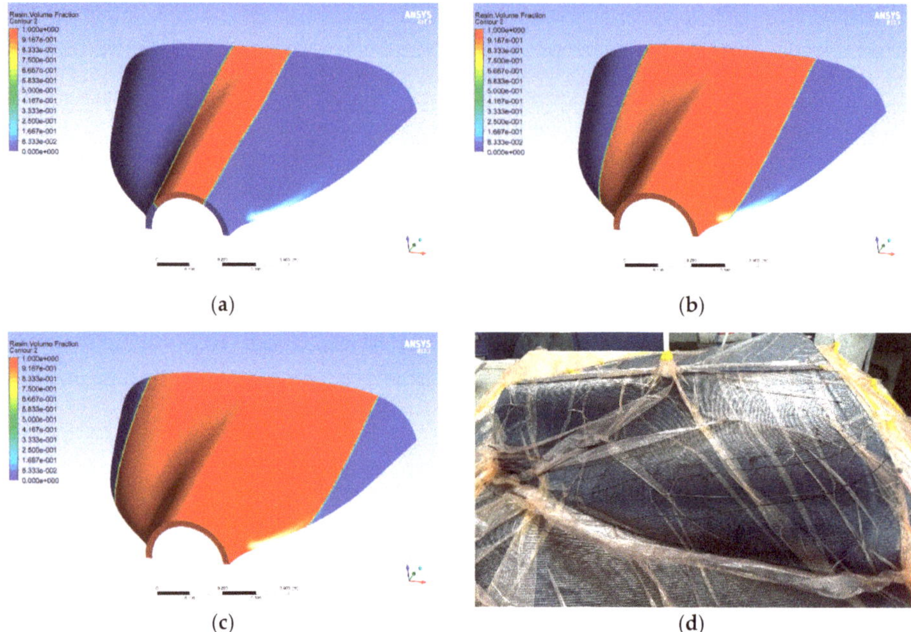

Figure 10. Results from the computational simulation of the resin impregnation process for the upper cowling using C-1 configuration: (**a**) t = 21 s; (**b**) t = 250 s; (**c**) t = 500 s and (**d**) resin infusion of cowling.

3.3. Verification of the Manufacturing Process of the Aileron Main Beam

For level 4 structural testing, mechanical tests of an aileron main beam were carried out. Master mold of aileron main beam was fabricated with MDF by CNC machining. After that a composite mold using twill carbon fiber and EPOLAM 2019 was fabricated. Manufacturing process of the aileron main beam follows the same VARI procedure described in Section 2.2. Layout of the beam is the same as described in Section 2.5. Post-curing for 2 h at 60 °C was the same as the sandwich coupons [17].

The manufacturing processes of the aileron main beam of the light sport aircraft were verified by comparison between the mechanical performances exhibited in testing and the results obtained from the FEM simulation. In Figure 11a, the experimental set up for the cantilever beam test of the aileron main beam is shown. The beam was fixed at one end and loaded at the opposite end. Load was applied manually by using sand bags. The beam was instrumented with two strain gauges at the fixed end in the upper and bottom surface respectively. Load was applied till the beam failure was reached.

The maximum deflection of the beam exhibited at the free edge in the experiment was 34 mm. The beam failed with a load of 35 kg. As can be observed in Figure 11b,c, the beam failed close to the fixed constraint. The results from the computational simulation showed a similar qualitative behavior respect the experimental test. The simulation was performed imposing a load of 343.35 N (35 kg) showing a maximum deformation of 35.02 mm (Figure 12a). With this deformation, the maximum value of the principal stress σ_1 is placed at the same position as the failure occurred during the mechanical testing (see Figure 12b). In Figure 13, the maximum strain failure criterion is plotted, as it is displayed, the maximum value of the failure criterion is 0.8437, indicating that the beam is able to support this load. However, the mechanical test showed the failure of the beam at these load conditions. The lack in the performance of the beam can be attributed to the intrinsic errors during the manufacturing process. As reported in literature [29], stiffened structure commonly fails at components interface. For woven composites, warp-weft sliding, fabric distortion and fiber decohesion are expected. In the case of sandwich structures, core crushing and compressed facing are also expected mainly for compression-tension stress state. For the aileron main beam, deflection due to the aerodynamic forces is expected, therefore, attention on fiber crushing and core crushing on the compression face shall be avoided.

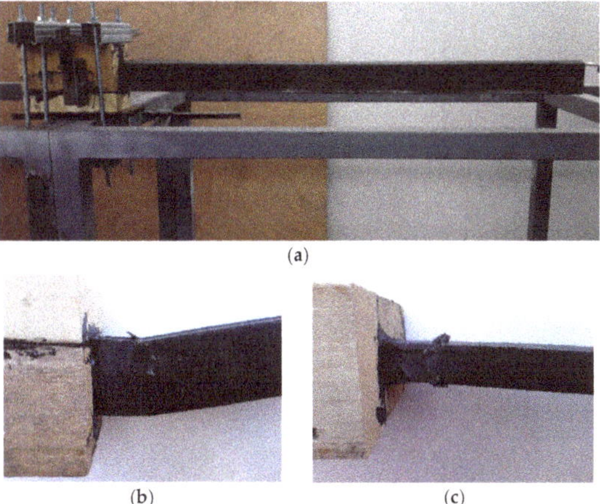

Figure 11. Mechanical test of the aileron main beam: (**a**) experimental set up; (**b**) front view and (**c**) top view of the beam failure.

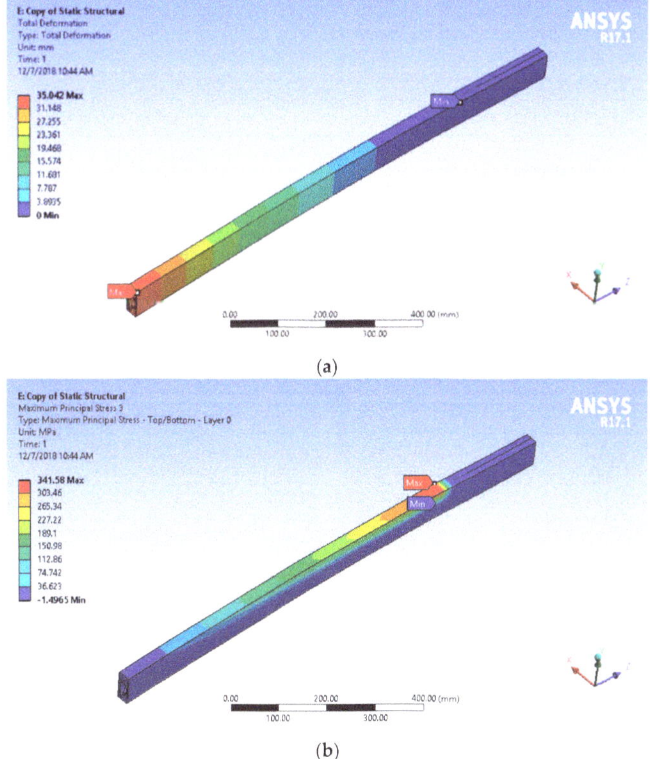

Figure 12. Finite Element Method (FEM) simulation results of the aileron beam in cantilever: (**a**) Total deformation; (**b**) Maximum principal stresses σ_1.

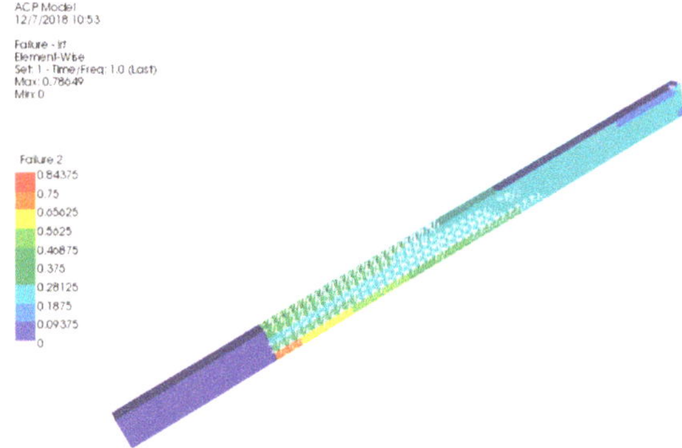

Figure 13. Maximum strain failure criterion for the composite beam with a load of 343.35 N.

In order to visualize in detail the internal structure of the beam and capture the manufacturing miscues, a computerized tomography analysis (CT-Scan) was carried out. In Figure 14b, the results from the three-dimensional reconstruction of the beam using the CT-Scan equipment are shown. A lateral and top slices at the middle planes of the beam structure are presented in Figure 14a. It can be seen the hexagonal PVC foam core on the left and the top and bottom fiber layers on the right side. The presence of pores is easily identified in the fibers and the hexagonal PVC foam core, the lack of resin in those zones can promote the premature failure of the beam with respect to the results obtained from the FEM simulation. In Figure 14c, a top slice at one of the UD fiber layer is shown, the pores and the fracture path in this fiber layer can be observed.

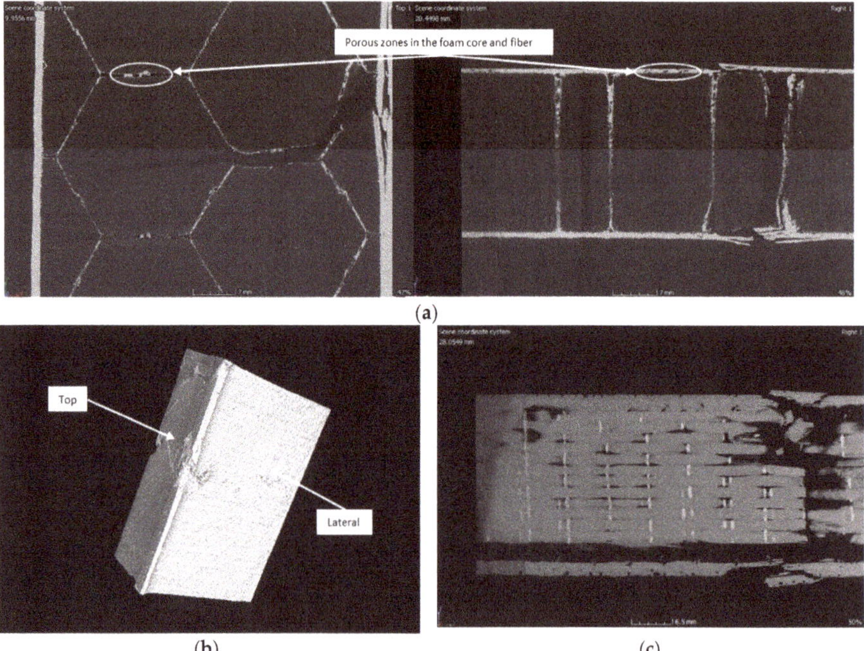

Figure 14. CT-Scan inspection of the beam after the mechanical test, (**a**) lateral and top slices inside the beam structure; (**b**) Three dimensional reconstruction of the beam specimen and (**c**) top slice of the laminate at the fiber layers of the beam flange.

4. Discussion and Conclusions

The manufacturing process development and validation for components of a light sport aircraft were presented. The VARI process was chosen to fabricate the composite components of the Horizontec aircraft to reduce the manufacturing cost versus an autoclave or RTM procedure [10,30,31]. The cost analysis of the VARI process was carried out by Horizontec when the project requirements were stablished. In order to show how the engineering and computational tools are utilized during the development of these manufacturing processes, two cases of study were presented. Firstly, the impregnation stage was analyzed for the upper cowling using CFD simulations. For this purpose, the permeability of the fibers was characterized to be used as input parameters of the simulation. CFD model was performed and the transient behavior of the resin flow during the filling process was computed. It is important to highlight that the CFD model was developed as simple as possible in order to reduce the computational time. It is expected to compute the global features of the flow and use these kind of CFD models as a fast engineering tool to design the impregnation strategy for each component of the aircraft. Here, three different inlet-outlet configurations to perform the impregnation process for the upper cowling were analyzed. From the simulations results, the filling time for each configuration was computed through the calculation of the resin volume fraction as a function of time. Then, the inlet-outlet configuration was selected in order to reduce the filling process. It is important to mention that with this model, it could be possible to propose a pressure or flow rate control methods to optimize the impregnation process. As a second case study, the mechanical testing and computational simulation for the aileron main beam was presented. Once the component is fabricated and in order to validate the manufacturing process, the component was tested and its mechanical performance was compared with results obtained from FEM simulation under the same load conditions. In order to develop robust simulations and model as well as possible the mechanical behavior of the aileron main beam before implementing the FEM simulation, the composite materials were characterized through coupons mechanical testing to feed the FEM simulation with the measured material properties. It was shown that for the case of the aileron beam manufactured by VARI process and in cantilever beam scenario, the beam has a performance of 85% with respect to the simulation results. Since all the mechanical properties for the corresponding materials were included into the FEM model and from the porosity zones identified from the CT-Scan analysis, it can be concluded that the 15% difference in performance is due to manufacturing errors.

For manufacturing LSA, the Building-Block Approach describes an excellent path for validating aircraft components. In both cases, cowling and aileron main beam, elastic properties of carbon fiber based structures were determined and fulfill the materials requirements according to LSA standards. The implementation of sandwich structures for both cases solves economic and engineering constraints. However, components must meet 1-to-1 scale mechanical tests to assure the certification type of Halcon 2 aircraft.

Author Contributions: Conceptualization, G.A. and M.T.; methodology, M.T., S.P., S.L. and C.A.E.-V.; software, S.P. and S.L.; validation, M.T. and S.L.; formal analysis, S.P., M.T. and S.L.; investigation, M.T., S.P., S.L. and C.A.E.-V.; resources, G.A. and M.T.; data curation, M.T. and C.A.E.-V.; Writing—Original Draft preparation, S.P., M.T. and S.L.; Writing—Review and Editing, M.T. and S.P.; visualization, S.P. and S.L.; supervision, G.A.; project administration, M.T. and G.A.; funding acquisition, G.A.

Funding: This research was funded by the Nacional Council of Science and Technology of Mexico (CONACYT) and the Ministry of Economy of Mexico (SE) through the Technological Innovation Fund (FIT), grant number ECO-2015-C01-260587. Experimental characterization was funded by CONACYT through the Research Fund for the Development of Space Activities (FIDAE), grant number 275783.

Acknowledgments: Mauricio Torres, Saul Piedra and Carlos A. Escalante-Velázquez convey their special appreciation to the "CONACYT Research Fellow Program (Cátedras CONACYT)." All of the authors convey their appreciation to Aaron Burgos, Miguel Vergara, Ricardo Lozada and Rodrigo Pérez for their technical support, as well as, Marco Paredes and Marc Emile Preud'homme from the CENTA NDT Laboratory.

Conflicts of Interest: The authors declare no conflict of interest.

References

1. DGAC CO AV-27/12. Mexican mandatory circular for technical requirement regulation for ultralight and experimental aircrafts to obtain an approval and technical requirement regulation for light sport aircraft to obtain a certification. In *Dirección General de Aeronáutica Civil*; SCT: Mexico City, Mexico, 2012.
2. ASTM F2245-16c. *Standard Specification for Design and Performance of a Light Sport Airplane*; American Society for Testing and Materials International: Philadelphia, PA, USA, 2016.
3. Baker, A.; Dutton, S.; Kelly, D. Aircraft applications and design issues. In *Composite Materials for Aircraft Structures*, 2nd ed.; Schetz, J.A., Ed.; AIAA Education Series: Washington, DC, USA, 2004; pp. 435–475, ISBN 1-56347-540-5.
4. Rouse, M.; Jegley, D.C.; McGowan, D.M.; Bush, H.G.; Waters, W.A. *Utilization of the Building-Block Approach in Structural Mechanics Research*; NASA: Greenbelt, CA, USA, 2005; pp. 105–116.
5. Park, J.; Ha, S.; Chang, F.K. Monitoring impact events using a system-identification method. *AIAA J.* **2009**, *47*, 25–36. [CrossRef]
6. McCarthy, C.T.; Gray, P.J. An analytical model for the prediction of load distribution in highly torque multi-bolt composite joints. *Compos. Struct.* **2010**, *93*, 17–27.
7. Crouzeix, L.; Davila, Y.; Collombet, F.; Douchin, B.; Grunevald, Y.H.; Zitoune, R.; Cénac, F. Study of double step lap composite repairs on CFRP evaluators. In Proceedings of the 15th European Conference on Composite Materials, Venice, Italy, 24–28 June 2012.
8. Lamkanfi, E.; Van Paepegem, W.; Degrieck, J. Shape optimization of a cruciform geometry for biaxial testing of polymers. *Polym. Test.* **2014**, *41*, 7–16. [CrossRef]
9. Torres, M.; Crouzeix, L.; Collombet, F.; Douchin, B.; Grunevald, Y.H. Numerical and experimental value added of multi-instrumented technological evaluator for the analysis of thick monolithic composite structures with singularity details. *Compos. Struct.* **2015**, *127*, 41–50. [CrossRef]
10. Meredith, J.; Bolson, E.; Powe, R.; Collings, E.; Kirwan, K. A performance versus cost analysis of prepreg carbon fiber epoxy energy absorption structures. *Compos. Struct.* **2015**, *124*, 206–213. [CrossRef]
11. Verma, K.K.; Dinesh, B.L.; Singh, K.; Gaddikeri, K.M.; Sundaram, R. Challenges in processing of a cocured wing test box using vacuum enhanced resin infusion technology (VERITy). *Procedia Mater. Sci.* **2014**, *6*, 331–340. [CrossRef]
12. Komarov, V.A.; Kurkin, E.I.; Spirina, M.O. Composite aerospace structures shape distortion during all stages of vacuum infusion production. *Procedia Eng.* **2017**, *185*, 39–145. [CrossRef]
13. Romano, F.; Barile, M.; Cacciapuoti, G.; Godard, J.L.; Vollaro, P.; Barabinot, P. Advanced OoA and automated technologies for manufacturing a composite outer wing box. *MATEC Web Conf.* **2018**, *5*, 1–8. [CrossRef]
14. Brouwer, W.D.; Van Herpt, E.C.; Labordus, M. Vacuum injection moulding for large structural applications. *Compos. Part A* **2003**, *34*, 551–558. [CrossRef]
15. HexTow®AS4 Carbon Fiber. In *Product Data Sheet*; Hexcel Corporation: Seguin, TX, USA, 2018; pp. 1–2.
16. HexTow®IM7 Carbon Fiber. In *Product Data Sheet*; Hexcel Corporation: Seguin, TX, USA, 2018; pp. 1–2.
17. EPOLAM 2019, Epoxy laminating system by infusion. In *Product Data Sheet*; Axson Technologies Head Office France: Saint Ouen L'Aumone, France, 2018; pp. 1–2.
18. PVC and XPS foam cores. In *Product Data Sheet*; 3D Core TM: Herford, Germany, 2018; pp. 1–2.
19. ASTM D3039. *Standard Test Method for Tensile Properties of Polymer Matrix Composite Materials*; ASTM International: West Conshohocken, PA, USA, 2014.
20. ASTM D3718. *Standard Test Method for In-Plane Shear Response of Polymer Matrix Composite Materials by Tensile Test of a ±45° Laminate*; ASTM International: West Conshohocken, PA, USA, 2013.
21. ASTM C393. *Standard Test Method for Core Shear Properties of Sandwich Constructions by Beam Flexure*; ASTM International: West Conshohocken, PA, USA, 2016.
22. Kang, M.K.; Jung, J.J.; Lee, W.I. Analysis of resin transfer moulding process with controlled multiple gates resin injection. *Compos. Part A* **2000**, *31*, 407–422. [CrossRef]
23. Trochu, F.; Ruiz, E.; Achim, V.; Soukane, S. Advanced numerical simulation of liquid composite molding for process analysis and optimization. *Compos. Part A* **2006**, *37*, 890–902. [CrossRef]
24. Laurenzi, S.; Grilli, A.; Pinna, M.; Nicola, F.; Cattaneo, G.; Marchetti, M. Process simulation for a large composite aeronautic beam by resin transfer molding. *Compos. Part B* **2014**, *57*, 47–55. [CrossRef]

25. Tryggvason, G.; Scardovelli, R.; Zaleski, S. *Direct Numerical Simulations of Gas-Liquid Multiphase Flows*; Cambridge University Press: Cambridge, UK, 2011; ISBN 9780511975264.

26. Escalante-Velazquez, C.A.; Piedra, S.; Jimenez, S.M.A.; Ledesma, S. Experimental characterization of the effect of a spray adhesive on the permeability of carbon fiber preforms utilizing the VARTM manufacturing technique. **2018**. Submitted.

27. Lundström, T.S.; Stenberg, R.; Bergström, R.; Partanen, H.; Birkeland, P.A. In-plane permeability measurements: A nordic round-robin study. *Compos. Part A* **2000**. [CrossRef]

28. Hoa, S.V. *Principles of the Manufacturing of Composite Materials*; DEStech Publications, Inc.: Lancaster, PA, USA, 2009; ISBN 9781932078268.

29. Torres, M.; Franco-Urquiza, E.A.; Hernández-Moreno, H.; González-Villa, M.A. Mechanical behavior of a fuselage stiffened carbon-epoxy panel under debonding load. *Aeronaut. Aerospace Eng.* **2018**, *7*, 2–7. [CrossRef]

30. Kaufmann, M.; Zenkert, D.; Mattei, C. Cost optimization of composite aircraft structures including variable laminate qualities. *Compos. Sci. Technol.* **2008**, *68*, 2748–2754. [CrossRef]

31. Witik, R.A.; Gaille, F.; Teuscher, R.; Ringwald, H.; Michaud, V.; Månson, J.A.E. Economic and environmental assessment of alternative production methods for composite aircraft components. *J. Clean. Prod.* **2012**, *30*, 91–102. [CrossRef]

Article

Manufacturing Aspects of Creating Low-Curvature Panels for Prospective Civil Aircraft

Evgeny Dubovikov *,[†] , Danil Fomin [†], Natalia Guseva [†], Ivan Kondakov [†], Evgeny Kruychkov [†], Ivan Mareskin [†] and Alexander Shanygin [†]

Central Aerohydrodynamic Institute (TsAGI), 1, Zhukovsky St., Zhukovsky, Moscow Region 140180, Russia; Danil.Fomin@tsagi.ru (D.F.); gus_nata2014@mail.ru (N.G.); Ivan.Kondakov@tsagi.ru (I.K.); Ivan.Mareskin@tsagi.ru (I.M.); Alexander.Shanygin@tsagi.ru (A.S.)

* Correspondence: evgeny.dubovikov@tsagi.ru

† These authors contributed equally to this work.

Received: 11 November 2018; Accepted: 30 January 2019; Published: 14 February 2019

Abstract: For this study, structural and manufacturing schemes for low-curvature pressurized fuselage panels were proposed, making it possible to provide high weight efficiency for the airframes of prospective civil blended wing-body (BWB) aircraft. The manufacturing scheme for low-curvature panels helped to achieve high strength characteristics of the composite details as well as decreased the labor input necessary for manufacturing and assembling. The beneficial features of the proposed structure are that the panels have a low weight, incur low manufacturing costs, and satisfy the demands of repairability.

Keywords: manufacturing; low-curvature panels; pressurized fuselage; blended wing-body aircraft

1. Introduction

Due to certain benefits, the blended wing-body (BWB) concept is considered to be one of the prospective concepts for the next generation of civil aircraft, especially for long-range airliners.

Estimating the effectiveness of applying the BWB concept to civil aircraft with high and extra high-passenger capacity has been the scope of several works undertaken by the Central Aerohydrodynamic Institute (TsAGI) within the framework of the European FP6 NACRE [1,2] project. Work under a number of contracts for Boeing and Airbus [3–5] has also been performed on this topic. In the course of that work, high aerodynamic characteristics and new feasible solutions concerning an aircraft's layout were obtained and substantiated. However, the question of the BWB concept's high weight efficiency remains because the application of conventional metallic and composite ("black metal") structural layouts do not provide an adequate weight efficiency. Consequently, the benefits related to the BWB concept, as compared to conventional analogues, have not been obtained.

The root of the problem of the BWB concept's low weight efficiency are the expenses related to high weight efficiency, namely, the costs needed to ensure the proper strength characteristics of high-loaded pressurized low-curvature (flat) panels. One solution to this problem was proposed in [6–8] based on double-lattice composite panels.

The main load-bearing elements of double-lattice pressurized panels are high-strength axisymmetric metal-composite rods that form a spatial frame. The frame, together with two composite lattice grids (inner and outer), is capable of effectively bearing both in-plane and out-of-plane loads (Figure 1). The panels also include low-loaded elements such as elastic skins (pressurized skin, shape-forming aerodynamic skin, and protective layers). These elements, despite their insignificant role in bearing mechanical loads, are important parts of the structure as they provide the high reliability of the structure, saving load-bearing elements from impacts and environmental factors.

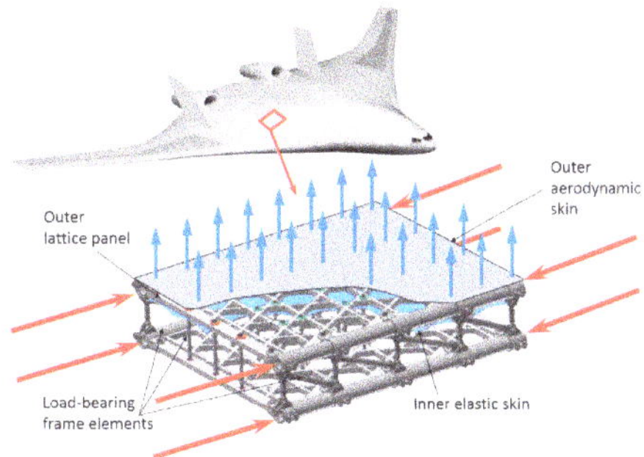

Figure 1. Principal scheme of bearing loads by the pressurized double-lattice flat panel.

In [6,7], it was shown that double-lattice pressurized panels, as described above, could create a low-weight airframe for the BWB with a high operational reliability. The application of such panels can therefore become the solid foundation for significant improvement in the transport efficiency of future aircraft by using the BWB concept.

For this work, the manufacturability of double-lattice panels was investigated and a manufacturing model was proposed and substantiated. In this paper, we present the results of the weight efficiency analysis of the double-lattice flat panels of a hypothetical BWB aircraft, taking into account the manufacturing aspects.

The main objective of the work was to show that the manufacturing requirements and constraints as well as the requirements related to operations and repairs do not drastically decrease the benefits related to the weight of this type of panel.

2. Critical Strength Tasks of BWB Airframes

Integral aircraft concepts, like the BWB concept, as applied to long-haul aircraft with high-passenger capacity, have a number of potential benefits when compared to conventional aircraft concepts [1,4,5]. The benefits of the BWB concept are created by higher aerodynamic characteristics and the possibility of using novel advanced layout solutions.

However, the BWB concept has a number of shortcomings caused mainly by two problems:

(1)   The large area of the outer surface, which leads to the appearance of large underloaded zones; and
(2)   The presence of large zones with a flat surface, subjected to considerable in-plane and out-of-plane loads.

This work focused on the second problem. The investigation centered on the upper flat pressurized panel, located in the most critical zone of the hypothetical civil BWB aircraft, where the mechanical and internal pressure loads are at a maximum. The panel covering this zone has the following dimensions: length $A = 3.8$ m, width $B = 3.84$ m, height $H = 0.42$ m.

The configuration and a list of the main parameters of the aircraft are presented in Figure 2. This hypothetical BWB aircraft is considered to be one of the prospective variants of the next generation of civil aircraft. Its maximal take-off weight equals 230,000 kg and its passenger capacity is 325 persons.

Parameter	Value
MTOW, kg	230000
PAX	325
Thrust, N	2x350000
Structure weight, kg / %	66150 / 28.76
Payload, kg	33500
Fuel, kg	70050
Aerodynamic quality	23.5
Specific fuel consumption, 10^{-4} kg/kg*s	1.44
Range, 10^3 m	12550

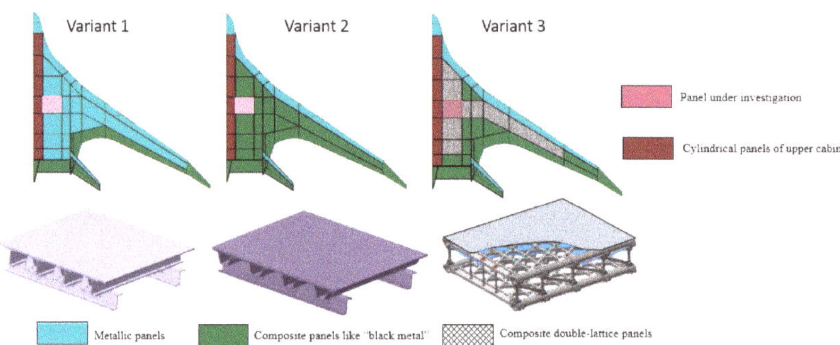

Figure 2. One of the prospective blended wing-body (BWB) configurations.

For this aircraft, the following three variants of pressurized panels with different layouts were investigated and compared: conventional metallic (variant 1), conventional composite (like "black metal") (variant 2), and double-lattice composite (variant 3) (Figure 3). For the first two variants, the weight characteristics were estimated by taking into account the layouts of the aircraft panels used in current civil aircraft. With respect to the third variant, the weight estimation was performed in parallel with the development of a manufacturing model of the double-lattice panel.

Figure 3. Alternative structure concepts for pressurized flat panels of the hypothetical BWB aircraft.

The weight of a square meter of the metallic panel was 70–75 kg, whereas the weight for the composite "black metal" variant of the panel was 65–70 kg. According to [9], the weight efficiency of a BWB structure turned out to be low, and practical possibilities to increase it significantly were not found.

In the present work, a comparative analysis of the weight characteristics of the above-mentioned panel variants was performed, taking into account the manufacturing aspects for the double-lattice panel.

3. Double-Lattice Pressurized Flat Panels

3.1. Specific Features of the Double-Lattice Concept

In a double-lattice pressurized flat panel, the main part of the primary structure is the spatial frame, consisting of axisymmetric rod elements. The other parts of the primary structure are two composite lattice grids (outer and inner). Additional elements are the elastic pressurized skin and the external shell (skin). The main structural elements of the double-lattice panel are represented in Figure 4.

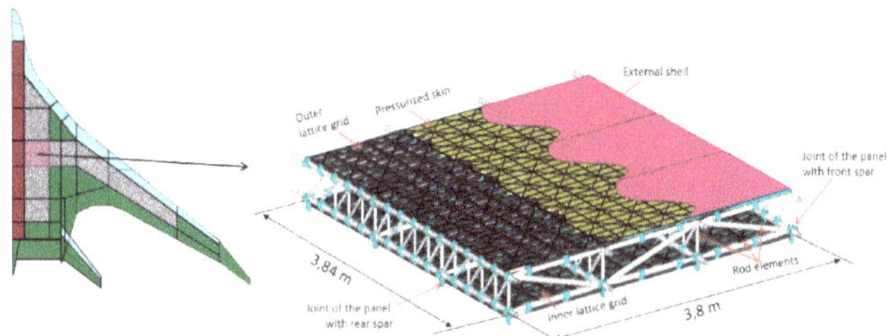

Figure 4. Double-lattice flat pressurized panel.

The main advantage of the double-lattice panels is related to its rational structural layout, allowing composite materials to work more effectively due to the unidirectional stacking of primary structural composite elements and their protection against impacts and environmental factors. Protective elements of the panel such as elastic skins and protection layers are used. Moreover, the lattice grids have high survivability due to their topology [10]. This makes it possible to significantly increase the allowable stress–strain level for primary structural elements of the flat pressurized panels and, as a consequence, increase their weight efficiency.

In the present work, a rational manufacturing concept for the double-lattice composite panel was investigated. For this purpose, the complex strength/weight numerical analysis and design of the upper panel of the hypothetical BWB's central pressurized wing box (Figure 4) were carried out. This analysis considered the factors of static loading, impacts, and environmental conditions.

Structural parameters of the double-lattice panel details were defined by taking into account the manufacturing constraints for the following composite manufacturing processes: fiber placement and wet winding. The design analysis was carried out using a four-level algorithm, developed at the TsAGI for the strength analysis and design of such metal-composite structures [6,11]. The algorithm was validated within a number of Russian (domestic) and European FP6-FP7 projects.

3.2. Four-Level Algorithm of Strength and Weight Analysis

The main specific feature of the four-level algorithm is that the design process is performed within a one complex iterative procedure including a number of global cycles. Within each global cycle, four sequential local actions are performed (Figure 5): aerodynamic analysis (level 1), structure layout analysis (level 2), manufacturing analysis (level 3), and strength analysis (level 4).

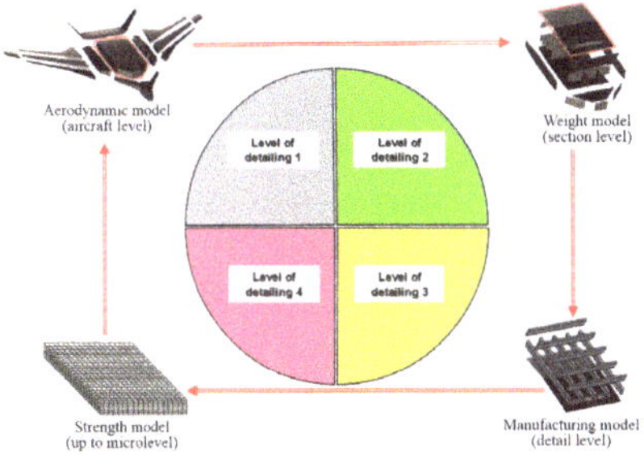

Figure 5. Scheme of the four-level algorithm.

To resolve the aerodynamic tasks (level 1), simplified finite element (FE) models are used to define the aerodynamic parameters and loads of the entire aircraft. The results obtained at this level are transferred to level 2 for the calculation of the structure parameters, weight characteristics, and aeroelastic characteristics. At this level, more detailed FE models are used for the airframe component. The obtained results are correspondingly transferred to level 3 for the definition of rational (feasible) manufacturing concepts. At this level, the manufacturability is checked and the manufacturing constraints are developed. FE models of level 3 are built for the details (subcomponents) of each component. Finally, at level 4, a strength analysis of the fragments/details is carried out.

Subsequently, the results of the global cycle are analyzed, the design parameters are changed, and the next iteration is launched. The iterative procedure stops when the values of the criteria become less than a defined value. As a rule, the main criterion is weight. More complicated criteria can also be used.

To significantly decrease the labor input and time needed for the calculations, the following principles were realized within the four-level algorithm [11]:

- Nesting principle for the FE models of all levels;
- Automated generation of auxiliary analytical models for each of the four FE models;
- Full automation of the iterative procedure;
- Application of specialized databases of prototypes for the weight analysis; and
- Application of specialized databases of available manufacturing processes for the definition of the manufacturing constraints.

4. Main Elements of the Double-Lattice Panel and Their Manufacturing Realization

4.1. Primary Structural Elements of the Flat Panel

4.1.1. Hybrid Rods

The main load-bearing elements of the frame structure of a double-lattice panel are hybrid (metal-composite) axisymmetric rod elements. The rod element (Figure 6) contains three parts: a cylindrical part and two conical parts. Additionally, the structural layout of such rods is also shown in Figure 6. The load-bearing part of the rod element is a composite shell with layers oriented close to its longitudinal axis. The inner space of the shell is filled with a lightweight material, which is used as a mandrel within the manufacturing (winding) process. The conical parts have metallic fittings at the ends. To save the weight, the metallic conical parts can have a porous structure with variable density, which could be maximal at zones of fittings, and considerably low at zones of attachments of metal and composite parts [12].

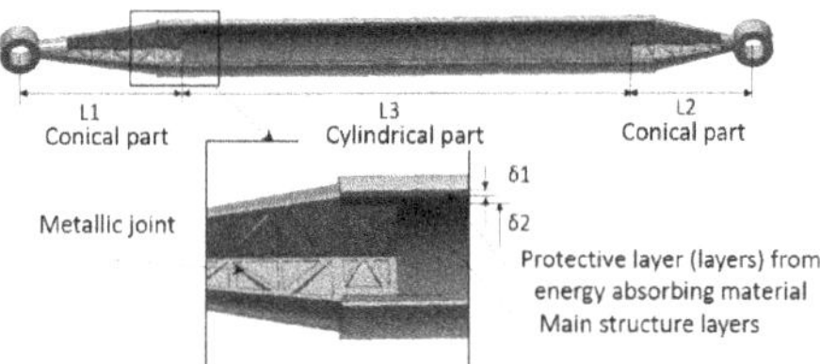

Figure 6. Main elements of the metal-composite rod.

Using a porous structure not only allows the ability to decrease the weight of the rod element, but can also balance the stiffness of the metallic and composite parts in zones of attachment [13].

The rod elements can sustain axial, torsion, and bending loads. However, they are most effective at longitudinal forces. The structure of the rod can have a special layer for impact and moisture protection, which can give a synergetic effect and also provide higher buckling margins of such rods. The manufacturing method for the rod elements combines winding for the composite parts and additive manufacturing (AM) for the metallic parts. It should be noted that for the length of a rod less than ~0.25 m and with a radius less than ~0.05 m, the metallic variant of the rod element is more effective.

At the first stage of the manufacturing process, the conical parts of the rod element are made using AM and are attached to the mandrel. At the second stage, winding is used for the final production of the rod element.

The assortment of the rod elements is defined by the following list of parameters:

L—length of the rod element;

$L1$, $L2$—length of the conical part;

$L3$—length of the cylindrical part;

$R0$—radius of the cylindrical part;

$R1$—radius of the conical part near fitting;

$\delta 1$—thickness of the protective layer;

$\delta 2$—thickness of the composite layer.

4.1.2. Main Joints

Three types of joints are used for the attachment of the rod elements: "end-to-end", "end-to-body", and "small end-to-body". First, two types are used to connect the rods, while the third is used to connect the rods with the lattice grids. The simplest variant of the joints, when only two rods are connected in one point, is shown on Figure 7. For the connection of more than two rods in one point, more complex joints are used. Connections of the rods with the lattice grid based on a "small-end-to-body" joint are shown in Figure 8.

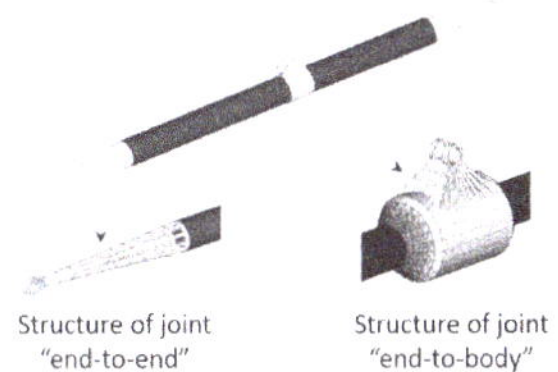

Figure 7. Examples of structural solutions for the attachment of hybrid rods.

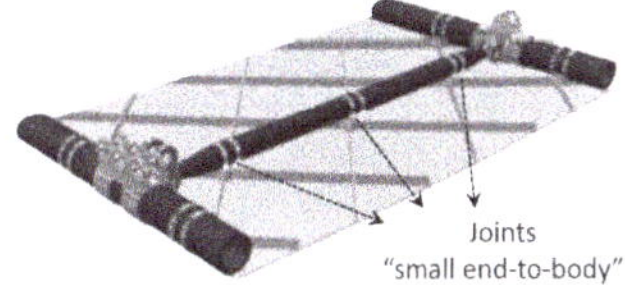

Figure 8. Example of the structural solution for the attachment of the aerodynamic skin and rod elements.

4.1.3. Lattice Composite Grids

The panel includes two lattice composite grids (outer and inner) with the same topology (Figure 9). The outer one is used for forming the aerodynamic shape and transferring aerodynamic loads to the frame, while the inner one is to sustain the internal pressure. The grids are attached to the load-bearing rods by means of "end-to-body" and "small end-to-body" joints using the attachment elements, and embedded into the grids. The number of such attachment elements is about 15–20 per square meter. The maximal load on a "small end-to-body" joint does not exceed 1 kN. For composite ribs of the inner grids, more elastic resins can be used, as the inner grid mostly sustains tension and bending.

For the outer grids, stronger and less elastic resins need to be used because the upper grid sustains bending and compression.

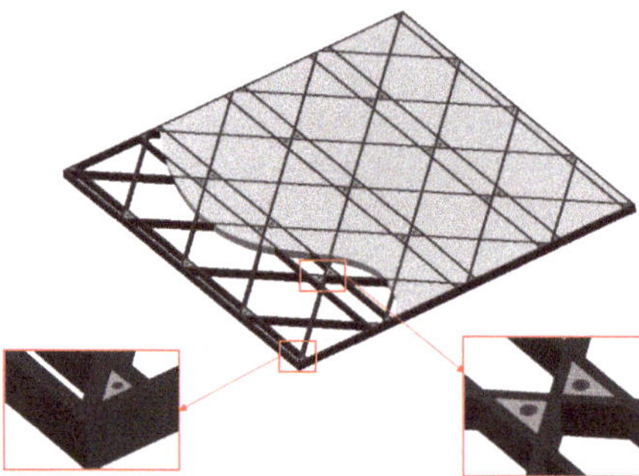

Figure 9. Lattice grid with a protective layer and embedded elements.

Manufacturing of the composite lattice grids can be performed using fiber placement or winding methods. In the frames of these conventional methods, the fibers are usually put into the grooves machined in the special foam coating. The foam coating is removed after curing. This manufacturing method was developed and validated in CRISM (Russia). For several high-loaded sections of the Russian Proton rockets, the weight benefits of lattice composite structures with respect to metallic prototypes were up to 25–50% [14].

In the manufacturing model proposed for the double-lattice panels, the foam coating is not removed as it serves as a protective layer. Such a manufacturing solution not only decreases the labor input and manufacturing costs and increases the quality of the grids (as the grid is not disturbed while removing the foam coatings), but also increases the strength characteristics of the panel.

The geometrical parameters and shapes for the inner and outer grids are different, but the pitch and angles of the orientation of the ribs for the grids are the same.

4.2. Low-Loaded Elements of the Flat Panel

4.2.1. Elastic Waveform Pressurized Skin

The structure of the double-lattice panel includes a thin elastic waveform pressurized skin. The pitch of the spherical segments of the skin coincide with the pitch of the ribs of the lattice grid. The wavelike form of the skin allows for a significant decrease in the tensile stresses in the skins caused by the internal pressure of 1 MPa, down to a level of 30–50 MPa, as the curvature radius of the "waves" is only 0.08–0.15 m (Figure 10).

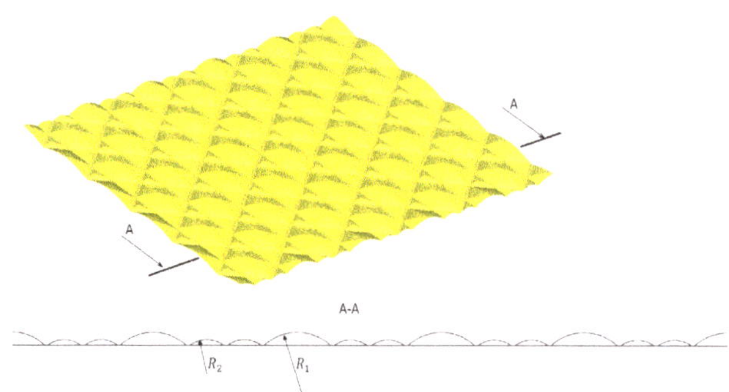

Figure 10. Waveform elastic skin.

As the pressurized skin only sustains tension loads, it is better to make it by means of braiding methods by using cheap low-strength elastic fibers and resins. Thermoplastic composites could be a good variant for such skins.

4.2.2. Outer Skin of the Panel

The panel includes an outer aerodynamic skin. The role of this element is to form the aerodynamic shape and transfer the aerodynamic loads to the outer lattice grid. It also serves as an impact-resistant protective layer. The skin is attached to the lattice grid by the means of bolts/pins.

The skin can be manufactured by means of fiber placement or braiding technologies. RTM-method (resin transfer molding) can also be applied.

Parameters of the skin are mostly defined by the stiffness requirements.

4.2.3. Thermal and Impact Protection Layers

Thermal/impact protection layers are the secondary parts of the panel and serve to:

- Form the local stiffness,
- Add thermal and impact protection,
- Attachment function (additionally, the outer skin can be connected to these layers by means of an adhesive joint),
- Foam coating for manufacturing.

4.3. Assortment of the Main Structural Elements

Table 1 shows a list of the main structural elements of the flat pressurized double-lattice panel and feasible variants of their manufacturing.

Table 1. Assortment of the main structure element of the panel.

Figure	Element	Material	Manufacturing Method
	Horizontal axisymmetric rods	Carbon fiber composite prepreg	Automated fiber placement/winding on special foam preform
	Transversal (diagonal) and vertical axisymmetric rods	Metal for ends and composite prepreg for tube	Automated fiber placement/winding on special foam preform
	Lattice grids	Carbon fiber composite prepreg	Automated fiber placement
	Joints for axisymmetric rods and lattices	Metallic alloy	Additive manufacturing (3D printing)
	Wave elastic pressurized skins	Fabric prepreg	Fiber placement /braiding
	Outer aerodynamic skins	Fabric prepreg	Fiber placement /braiding
	Wave foam covering	Foam	Cutting
	Attachments (pins, bolts, screws)	Metallic alloy	Conventional metallic technologies

5. Multilevel Principle of Manufacturing and Repair of the Flat Double-Lattice Panel

Three following main units can be distinguished in the structure of the double-lattice panel:

1) Spatial frame skeleton including the rod elements and joints;
2) Outer subpanel including the outer lattice grid, wave elastic hermetic skin, wave foam covering, foam covering with embedded elements and outer elastic skin; and
3) Inner subpanel including the inner lattice grid, flat foam covering with embedded elements.

5.1. Main Stages of Manufacturing of the Panel

The double-lattice panel is manufactured in three main stages (levels). At the first stage, the foam coverings for the rod elements and the subpanels (Figure 11) are manufactured by means of turning and cutting. In parallel, the metallic conical parts for the rod elements and other attachment elements are manufactured using AM and conventional methods.

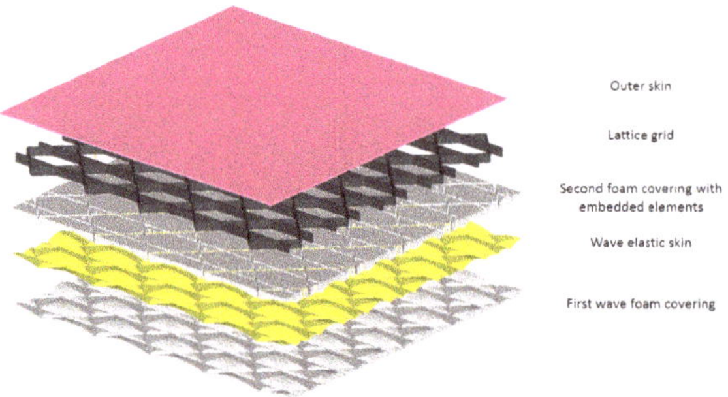

Figure 11. Main structural elements of the upper subpanel of double-lattice panel.

At the second stage, the rod elements and the subpanels are manufactured using wet winding and fiber placement. The second stage contains a number of substages for manufacturing both the rod elements and the lattice subpanels.

With regard to the rod elements, there are three substages. During the first substage, a cylindrical foam covering is connected with metallic conical parts. During the second substage, a number of layers (including load-bearing and protection layers) are winded. During the third substage, the rod elements are cured. This manufacturing process of the second stage can be automated to a high extent.

For the outer lattice subpanel, there are three substages. During the first substage, a wave elastic skin is manufactured by means of fiber placement using the first wave foam covering. At the second substage, the second foam covering is put on the wave skin and pressed. Next, the ribs of the lattice grid are placed into the grooves of the second foam covering and the attachment elements are embedded. At the third substage, the outer skin is placed and connected to the subpanel using bolts/pins and adhesive joints. Finally, the subpanel is cured.

For the inner lattice subpanel, which has no skin, only one substage is needed. Within this substage, the lattice grid is placed into the grooves of the foam covering and is cured.

At the third stage, the double-lattice panel is finally assembled in three substages. At the first substage, the spatial frame is assembled from rod elements using "end-to-end" and "end-to-body" joints. At the second substage, the outer lattice subpanel is connected to the frame by means of bolted joints. At the third substage, the inner lattice subpanel is attached.

5.2. Multilevel Protection System for Primary Structural Elements of the Panel

The sensitivity of the primary structural composite elements to impacts and environmental damage significantly decreases their weight efficiency. One of the methods against impact damage is through the protection of the primary structural elements from impacts by means of the application of special impact-absorbing materials.

Investigations dedicated to the development of special protection for unidirectional composite high-loaded rod elements were carried out in TsAGI [15].

For a high-loaded rod element, a special protection system against impacts and environmental factors was proposed that consisted of a combination of protective layers winded on the rod element (Figure 12):

- Layer of carbon fibers,
- Layer of elastic impact-absorbing materials,
- Layer of aramid fabric, distributing the impact energy on a larger area,
- Layer of chemical-resistant material.

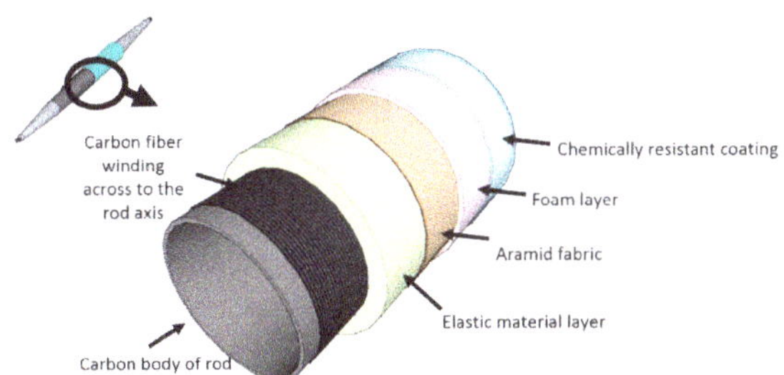

Figure 12. Concept of the special protection system for rod elements against impacts and environmental factors.

The materials used for these layers had low moisture absorption, high impact resistance, and high energy absorbing properties. The results of the investigations also showed that increasing the strength of high-loaded composite rod elements was more effective when adding material to the protection system, instead of adding the material to the element itself, by taking into account the degradation of its properties caused by the environment and impacts (see Figure 13).

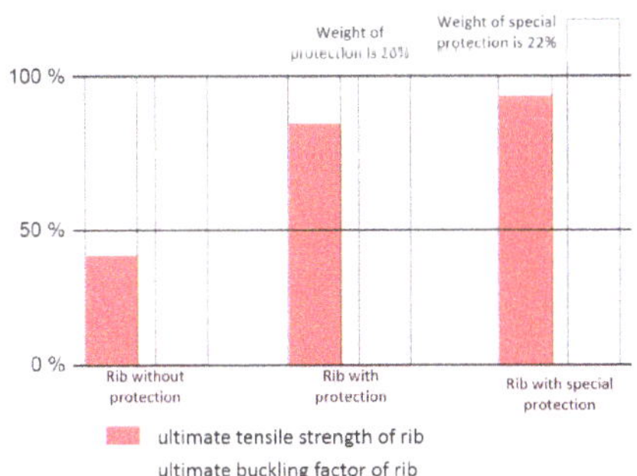

Figure 13. Influence of different variants of the protection on the residual strength and buckling properties of the composite rod elements.

Figure 14 shows the diagrams characterizing the strength parameters of cylindrical rods with various extents of protection after impact loading with energy ~20 J. It is shown, that spending ~20% of weight of the rod element on its protection makes it possible to not only fully maintain its strength characteristics after 20 J of impact, but even increase them due to synergetic effects. In particular, using the special multilayered protection (Figure 12) with a weight of 22% of the rib weight increased the buckling factor of the rib by ~5% due to squeeze caused by the wrapping layer of carbon fibers. Such results were also obtained for the lattice grid.

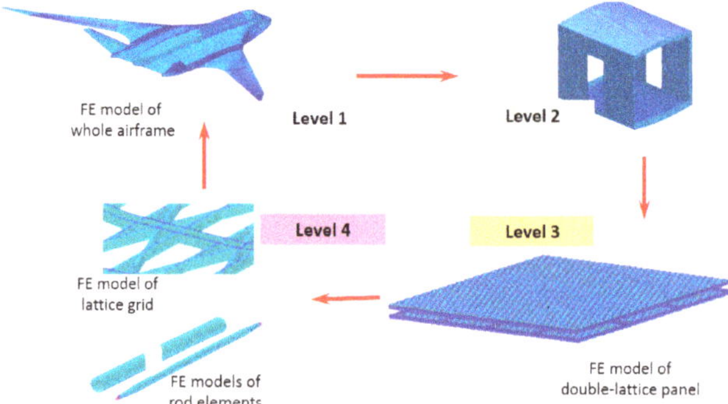

Figure 14. FE models of different levels for the weight estimation of the double-lattice panel.

Another method of decreasing the influence of impact loads on the performance of the structural elements is the location of the element inside the structure and the creation of a "natural" protective system for it. The "natural" protective system consists of both impact-absorbing layers and structural elements, naturally protecting this element from impacts.

In the structure of the pressurized double-lattice panel, both of the above-mentioned methods are used. For example, the main elements of the panel, the rod elements of the frame, are protected against external impacts by means of an aerodynamic skin with an impact-protection layer, upper lattice subpanel, and the rod elements of the upper frame. By this reason, there is no need to create a special protection system for these elements.

The lattice grid of the outer subpanels is protected from impacts, correspondingly, by means of an aerodynamic skin with an impact-protection layer. The lattice grid has a multi-path layout, so the impact damage is less dangerous for its load-bearing capacity.

Finally, the aerodynamic skin is subjected to direct impact and can be damaged. However, the damage will be even less critical than the damage of the lattice grid as the skin is not a primary structural element.

For the double-lattice panel, a simple and cheap enough repair scenario can be realized. Three types of repair of the panel, depending on the level of damage, can be classified:

1. Minor damage repair: Repair of the skin and several ribs of the lattice grid. This is an in field type of repair and only requires standard equipment. The damaged skin is repaired using patches by means of bonding. The damaged zone of the lattice grid is filled with a polymer compound.
2. Medium damage repair: Replacement of the skin and lattice for a significant part of the panel. Such types of repairs presume the replacement of the upper lattice sub-panel with a new one. The lattice sub-panel is connected with the frame by bolts and so can easily be removed.
3. Hard damage repair: Replacement of the skin and one to two elements of the upper frame.

In this case, the upper lattice sup-panel is removed to provide access to the frame. After that, the damaged rods are repaired or replaced. The upper lattice sub-panel is also replaced.

Such repair scenarios can provide for the long-term and reliable operation of the structure with low weight and cost expenses.

6. Estimation of Weight Efficiency of the Flat Pressurized Panel

In this work, the weight analysis of alternative variants of the pressurized panels including the double-lattice panel structure was performed using the four-level algorithm. For all variants,

FE models, corresponding to each level of the algorithm, were built, taking into account the special features of each layout. The types of models for levels 1 and 2 were the same for each of the variants, while for levels 3 and 4, different types of models were different. For variants 1 and 2, conventional FE models based on 2D shell elements for the panel (level 3) and its fragments including stiffeners (level 4) were built. Regarding variant 3, special FE models of levels 3 and 4, forming lattice grids and axisymmetric rod elements, were built (Figure 14).

All FE models for all variants were fully automated and the data transfer between the levels during the cyclic procedure of the design analysis was organized without manual operations. In the design analysis, all elements of the BWB aircraft structure were designed and analyzed. However, the comparative weight analysis was performed only for the above-mentioned pressurized panel.

Parameters of the structural materials, used in the FE models, are shown in Tables 2 and 3.

Table 2. Parameters of the structural materials for variants 1 and 2.

Element	Metallic				Black-metal Composite			
	Elastic modulus, GPa	Allowable stresses, MPa			Longitudinal elastic modulus, GPa	Allowable strains		
		Tension	Compression	Shear		Tension	Compression	Shear
Skin	72	350	250	200	57	0.4%	0.3%	0.5%
Stringers/ frames	71	350	250		68.67	0.4%	0.3%	0.5%
Webs of spars/ribs	71			200	68.67	0.3%	0.2%	0.23%

Table 3. Parameters of the structural materials for the double-lattice panel.

Element	Material Name	Density kg/m³	Longitudinal Elastic/ Shear Modulus GPa	Allowable Strains/Stresses		
				Tension	Compression	Shear
Lattice grids	Hexply M21	1600	90/5.3	0.6%	0.5%	
Composite parts of rod elements	Hexply M21	1600	88.6/8.7	0.8%	0.7%	
Inner skin	Polystyrene	1050	1.3/-	80 MPa		
Outer skin	Hexply M21	1600	57/21.5	0.4%	0.3%	0.2%
Joints	VT6 titanium alloy	4500	105/40.38	900 MPa	900 MPa	600 MPa

Critical loads for the panel were derived on the basis of the FE and analytical models of levels 1 and 2, while detailed manufacturing and the strength analysis of the panel were carried out using FE models of levels 3 and 4. The main design load case for the upper panels was the maneuver case with a load factor of 2.5 and internal pressure of 1 atm. The value of maximal bending moment in the zone of the pressurized cabin (containing the upper panel) was $M_{bend} \approx 25 \times 10^6$ N·m, torsion moment $M_{tors} \approx 3 \times 10^6$ N·m. The value of the normal force for the upper panel was $P_{norm} \approx 7.9 \times 10^6$ N. These load factors are sustained mainly by the longitudinal (along spanwise) rod elements. The lattice grids of the upper panel bear the local loads from the internal pressure and the aerodynamic forces and transfer it to the longitudinal rod elements.

The results of the numerical strength analysis of the panel showed that the values of the structural parameters (Table 4), obtained during the design analysis, satisfied both the strength requirements and the constraints on the out-of-plane displacements of the panel, defined by the aerodynamic constraints (limits for the drag coefficients).

Table 5 shows the values of the weights for the main structural elements for one square meter of the upper panel for all variants of the structure.

Table 4. Rational values of the structural parameters.

Lattice Grids (Outer and Inner)		Rod Elements	
Parameter	Value, m	Parameter	Value, m
Height of ribs	0.03	Outer diameter of the inner horizontal axisymmetric rods	0.1
Width of diagonal ribs	0.006	Thickness of wall of the inner horizontal axisymmetric rods	0.017
Width of perimeter ribs	0.009	Outer diameter of the outer horizontal axisymmetric rods	0.076
Width of ribs orthogonal	0.0025	Thickness of wall of the outer horizontal axisymmetric rods	0.011
		Outer diameter of the transversal (diagonal) and vertical axisymmetric rods	0.02
		Thickness of wall of the transversal (diagonal) and vertical axisymmetric rods	0.003

Table 5. Results of the weight analysis: specific weights of the components of the panels, kg/m^2.

Metallic Panel Elements	Weight	Composite Panel Elements	Weight	Double-lattice Panel Elements	Weight
Skin	23.7	Skin	24.3	Frame grid	28.7
Longitudinal stiffeners	18.2	Longitudinal stiffeners	16.6	Inner lattice grid with attachment	9.2
Transversal stiffeners	12.1	Transversal stiffeners	11.9	Outer lattice grid with attachment	10.0
Flanges of ribs	4.2	Flanges of ribs	3.9	Inner pressurized skin	0.6
Flanges of spars	4.9	Flanges of spars	4.5	Outer skin with thermal insulation layer	2.2
Attachment elements	5.3	Attachment elements	4.0	Protection of primary structural elements (including foam coverings)	3.0
				Additional attachment elements	2.4
Total	68.4 (100%)		65.2 (95%)		56.1 (82%)

7. Conclusions

The proposed concept of a flat double-lattice panel of the pressurized cabin of the BWB aircraft was analyzed from the viewpoint of the feasible manufacturing methods using currently available processes of aircraft production. A rational and simple manufacturing model for the double-lattice pressurized panel was proposed. Weight analysis showed that the manufacturing demands did not disturb the potential benefits of the double-lattice pressurized panel in weight saving. The weight comparison of the alternative variants made it possible to substantiate up to 18% weight saving of the double-lattice panels in comparison with the metallic analogue and up to 14% in comparison with the conventional composite analogue.

An effective system of multilevel protection and simple and reliable repair scenarios were also substantiated for the double-lattice panel.

The next step of the investigations will be manufacturing and testing the demonstrator of the double-lattice panel.

Author Contributions: Investigation, E.D., D.F., I.K., E.K., I.M., and A.S.; Writing—original draft preparation, E.D., D.F., N.G., I.K., I.M., and A.S.; Writing—review and editing, A.S.; Supervision, A.S.

Funding: This research was funded by the Ministry of Industry and Trade of Russia in the framework of the State program on the development of the civil aviation industry.

Conflicts of Interest: The authors declare no conflict of interest.

References

1. FP6 NACRE Project. Available online: https://cordis.europa.eu/project/rcn/75773/reporting/en. (accessed on 13 February 2019).
2. Okonkwo, P.; Smith, H. Review of evolving trends in blended wing body aircraft design. *Prog. Aerosp. Sci.* **2016**, *82*, 1–23. [CrossRef]
3. Investigations of Technologies, Critical for Implementing an Airplane of Flying Wing Type with Superhigh Seating Capacity, ISTC Project #0548. Available online: http://www.istc.int/en/project/CAA832BC090C56E44325691F0011CAD4 (accessed on 13 February 2019).
4. Denisov, V.E.; Bolsunovsky, A.L.; Buzoverya, N.P.; Gurevich, B.I.; Shkadov, L.M. Conceptual Design for Passenger Airplane of Very Large Passenger Capacity in the Flying Wing Layout. In Proceedings of the 20th Congress of the International Council of the Aeronautical Sciences, Sorrento, Italy, 8–13 September 1996; Volume II, pp. 1305–1311.
5. Bolsunovsky, A.L.; Buzoverya, N.P.; Gurevich, B.I.; Denisov, V.E.; Shkadov, L.M.; Udzhukhu, A.Y. Investigations on possible characteristics of FW superhigh seating capacity airplane. In Proceedings of the 22nd Congress of the International Council of the Aeronautical Sciences, Harrogate, UK, 27 August–1 September 2000; pp. 116.1–116.9.
6. Shanygin, A.; Dubovikov, E.; Fomin, V.; Mareskin, I.; Zichenkov, M. Designing pro-composite truss layout for load-bearing aircraft structures. *Fatigue Fract. Eng. Mater. Struct.* **2017**, *40*, 1612–1623. [CrossRef]
7. Mareskin, I. Multidisciplinary optimization of pro-composite structural layouts of high loaded aircraft constructions. In Proceedings of the 30th Congress of the International Council of the Aeronautical Sciences, Daejeon, Korea, 25–30 September 2016.
8. Dubovikov, E.; Fomin, V.; Kondakov, I.; Shanygin, A.; Vedernikov, D. Development of rational hybrid fuselage structure for prospective civil aircraft. *Proc. Inst. Mech. Eng. Part G J. Aerosp. Eng.* **2018**. [CrossRef]
9. Dubovikov, E.A. Novel approach and algorithm for searching rational nonconventional airframe concepts of new generation aircrafts. In Proceedings of the 28th Congress of the International Council of the Aeronautical Sciences (ICAS 2012), Brisbane, Australia, 23–28 September 2012.
10. FP7 PoLaRBEAR Project. Available online: http://cordis.europa.eu/result/rcn/175990_en.html (accessed on 13 February 2019).
11. Shanygin, A.; Fomin, V.; Zamula, G. Multilevel approach for strength and weight analyses of composite airframe structures. In Proceedings of the 27th Congress of the International Council of the Aeronautical Sciences, Nice, France, 19–24 September 2010.
12. Kondakov, I.; Mareskin, I.; Shanygin, A.; Zichenkov, M. Features of design of hybrid airframes on the basis of unidirectional composite elements and metallic metamaterials. In Proceedings of the 15th Annual Russian-French Seminar TsAGI-ONERA, Sochi, Russia, 4–7 October 2016.
13. Chernov, A.; Fomin, D.; Kondakov, I.; Mareskin, I.; Shanygin, A. Lightweight and reliable metal-composite joints based on harmonization of strength properties of joined parts. *Proc. Inst. Mech. Eng. Part G J. Aerosp. Eng.* **2018**, *232*, 2663–2672. [CrossRef]
14. Vasiliev, V.V.; Razin, A.F. Anisogrid composite lattice structures for spacecraft and aircraft applications. *Compos. Struct.* **2006**, *76*, 182–189. [CrossRef]
15. Dubovikov, E.; Fomin, V.; Hühne, C.; Kolesnikov, B.; Kondakov, I.; Niemann, S.; Shanygin, A.; Wagner, R. Anisogrid design for fuselage primary structures—Results of EU/RU projects ALaSCA and PoLaRBEAR. In Proceedings of the Royal Aeronautical Society's 5th Aircraft Structural Design Conference, Manchester, UK, 4–6 October 2016.

MDPI

St. Alban-Anlage 66

4052 Basel

Switzerland

Tel. +41 61 683 77 34

Fax +41 61 302 89 18

www.mdpi.com

Aerospace Editorial Office

E-mail: aerospace@mdpi.com

www.mdpi.com/journal/aerospace